MAMMALS
IN
CARMARTHENSHIRE

MAMALIAID YN
SIR GAERFYRDDIN

Andrew Lucas

Cover by James Field
Illustrations by Tom McOwat

Published by Andrew Lucas
Printed by Gwasg Gomer, Llandysul, Ceredigion, Wales.

ISBN 0 9532112 0 7

Mammals in Carmarthenshire was supported by

CYNGOR CEFN GWLAD CYMRU
COUNTRYSIDE COUNCIL FOR WALES

The Mammal Society

Pryce Consultant Ecologists

Mammals in Carmarthenshire is dedicated to the memory of George Mainwaring. A good friend, whose conversation and encouragement are greatly missed.

FOREWORD

For those of us who rely on the skill, knowledge and hard work of others to provide the enjoyment of owning and learning from books on British natural history, it is always a great pleasure to welcome a new volume on the subject.

The wilderness that was once all Wales is gone. Nevertheless, her shapely unyielding body and her unforgiving weather have made her a natural fortress. A refuge for wildlife, without which the whole of southern Britain would now be much the poorer.

Mammals in Carmarthenshire, is therefore doubly welcome, because it is about the natural history of Wales, and deals with a group which are rarely seen and insufficiently understood.

For most of us, our encounters with live mammals in the wild, are those all too brief, glimpses of something which might have been. Within these pages you will surely find what it could have been. Unless of course it's a new record.

In bringing together, and adding to, that which is known about the mammals of Carmarthenshire, Andrew Lucas has enriched the lives of all of us who recognise the joy and importance of an understanding of wild things.

Lionel Kelleway
Carmarthenshire,
November 1997

ACKNOWLEDGEMENTS

It quickly became apparent during the research for *Mammals in Carmarthenshire* that I was, as the phrase goes, 'putting my sickle into other peoples corn'. I have been assisted by numerous specialists whose generous help I am only too happy to acknowledge.

Ian Morgan commented on at least three 'first' drafts. Few people can know more about Carmarthenshire wildlife than he does and his knowledge and encouragement are greatly appreciated. Ian also contributed many of the photographs. Richard Pryce's rigorous eye for detail eliminated many typographical errors and helped to contain my sometimes rambling English. In his role as BSBI recorder Richard also allowed me to use the whorled caraway distribution map. The text also benefited from the ruthless editing of David Saunders and Annie Poole, which added greatly to its clarity.

Tom McOwat gave freely of his expertise in desk-top publishing, without which *Mammals in Carmarthenshire* would never have made it off the word processor, as well as providing drawings for the text and contributing much of his encyclopaedic knowledge of bats in Carmarthenshire.

Particular thanks should go to the previous mammal recorders for Carmarthenshire, Shelagh Seery and Richard Rees, without whose dedicated efforts the distribution maps would have been far less informative.

I would like to thank the following people who contributed their memories of Carmarthenshire wildlife; Fred Bellamy, Tom and Gwen Heal, Jan and Keith Crowden, Lord Dynevor, Neil Mathew, Caradog Jones, Barry Penny, Alan Enoch, Mike Scott, John Humphries, Dafydd Davies and Stan Rudge. Particular thanks are due to Mr Ifor Evans of Tupsley, Hereford who, unprompted, contributed his recollections of wildlife in the Llandybie area in the 1930s. Gillie Sargent and Pat Morris offered advice from a national perspective.

Mike Sharrot advised on the content of the badger species account. Heather James of Cambria Archaeological Society assisted with many historical details, as well as allowing me to use the CAS place name database to search for mammal names. Peter Howlett of the National Museum of Wales made much archive material available, as well as helping me trawl through the Museum's extensive skin collection.

Special thanks are due to John Messenger, Johnny Birks and Don Jeffries who kindly made available their data from the Vincent Wildlife Trust pine marten survey ahead of publication. David Wooley provided data from the Dyfed Bat Group database.

6The maps were constructed from records entered onto the BIORECS database designed by Steve Coker, using the DMAP mapping package written by Dr Alan Morton. Thanks to both for their assistance.

The cooperation of Dr David George of the Marine Biological Services Division of the Natural History Museum in making cetacean records available is gratefully acknowledged.

Thank you also to Dic Davies of Llanelli Library who kept an eye out for unusual mammal stories, and unearthed, amongst others, the articles on the Llangennech Whale and the Pinged Otters.

The financial support of the Countryside Council for Wales, the Mammal Society and Pryce Ecological Consultants is greatly appreciated.

Mammals in Carmarthenshire could not have been written without the numerous people who have contributed records, both to me and to my predecessors over the years. My sincere thanks to them all:

Mrs Aldrige, Mrs Armstrong-Esther, Mr Beresford, Mr and Mrs Giffin, Mr Harrison-Rees, Mr and Mrs May, Tony Braithwaite, Alison Brown, Alan Davies, A Deacon, Alan Enoch, Tony Jenkins, A Richards, A Vionee, A Waddington, Alison Wheeler, B Davies, B Fox, Pat and Barney Gill, Barry Long, B Reeby, Barry Stewart, B Stringer, Carolyn Hunt, C Jones, C Masters, C Nowell, C Richardson, Christine Rosters, D Brown, D Cadlow, D Davies, Dawn Gray, David Herbert, DS Hughes, D Jarman, Dave Lewis, D Long, D Pajak, David Poulter, D Pugh, DR Rees, D Roberts, David Smith, David Stacey, D Witchbold, Dyfed Bat Group, Dewi Davies, E Lewis, Eira Prior, EE Roderick, EJ Smith, EB Thomas, F Emmet, Fiona Lanc, Fiona Macpherson, F Merrett, GC Cundale, Graham Clarke, G Cundiff, GA David, Gerry Dunn, Gavin Hall, George Hutchinson, Graham Motley, George Noakes, Hilary Cackett, Hazel Drewett, H Higgs, Howard Jones, Hywel Manley, Harry Spiller, Huw Williams, I Evans, Ian Hainsworth, I Rolls, Miss J Benham, Jamie Bevan, John Bezzant, J Branscombe, Mrs J Buchanan, John Comont, Jonathan Copp, Mrs Jenny Crane, Jim Davies, J Donovan, J Friese, John Galvin, Jonathan Graham, Jean Hemmings, J Hiff, John Hodgson, Jill Ilsley, Jackie Inskip, John Judd, John Lloyd, James McCallum, Jill Martin, Judith Phillips, Mr J Poulson, Jackie Sampey, John Shrimpton, John Steer, Mrs J Williams, Kevin Clements, Ken Davies, Dr K Lewis, Kath Miller, KT Morgan, Kyra Somerfield, Keith Williams, LE Davies, Leo Evans, Lin Gander, Louise Humphrey, Lionel Kelleway, Louise Liles, Mick Baines, Marjorie Clark, Marion Herbert, M Higgs, MG Hill, Dr Malcolm Holding, Martin Humphrey, Max James, Morfydd Lloyd, ML Morris, M Nichols, Maggs Richards, Matt Ridley, M Roberts, M Rowlands, Mike Scott, Mary Thomas, Mike Wade, M and J Iliff, MH Williams, Mrs Wrtz, N Fox, Nigel Stringer, Nic Thomas, ND Vessey, O Daniels, P Beeson, Peggy Blogg, P Chasty, PC Cox, Phil Morgan, Pam Morris, P Stoney, Quentin MacGarvie, Richard Davies, Rob Donaldson, Richard Edwards, Ray Garner, Richard Howarth,

RAD Hughes, P Nathan, Miss R Powell, R Rees, Rosemary Regelous, Richard Smale, Robert Wynne-Davies, R Whitehead, Sarah Andrews, S Davies, Stephen Evans, Sally Ford, Simon Good, Steve Lucas, Sandra Ozollins, Steve Pocock, Dr S Rhodes, S Roberts, Sally Thornton, Steve Thomas, Tom Hallawell, Tim Jones, T Morrisey, Tony Pickup, T Wells, W Eardley, W Harcup, Gelli Aur WATCH, Welsh Office Agricultural Department, Sheila Lewis, Mal Jones and Mollie Mizen.

I apologise to any people who, through error, I have omitted from this list.

Thanks are also due to my wife, Beth, who encouraged me when I thought I should forget the whole thing, and without whose word processing expertise *Mammals in Carmarthenshire* could never have been written. Both Beth and my daughter Rhiannon put up with me disappearing at all hours of the day and night to sit in front of the computer, particularly in late 1997 when work had to be completed quickly.

Finally, I would like to thank my parents, Bob and Joyce Lucas, who ignored my headmaster's advice and encouraged my interest in natural history.

A NOTE ON PLACE NAMES

Place names may seem an esoteric point on which to start a book about mammals. However, with many places in Carmarthenshire having two names, both Welsh and English, this can be a contentious subject.

Nothing infuriates the author more than the casual and rather lazy corruption of beautiful Welsh names. However, any name is useless if it does not convey its location clearly to readers, both in Wales and beyond.

Therefore, with apologies to those who prefer Talyllychau to Talley, or Cydweli to Kidwelly, I have decided to use those names currently in use on the 1:50 000 Ordnance Survey maps of the area. I have allowed only two exceptions to this rule. Firstly, Mynydd Du is used in preference to Black Mountain, in order to differentiate between it and the Black Mountains in the eastern part of the Brecon Beacons. Secondly, Llansteffan is preferred in place of Llanstephan, the former being universally used on road signs and closer to the usual pronunciation.

National grid references for all Carmarthenshire locations mentioned in the text are given in Appendix 4.

CONTENTS

CHAPTER 1:
INTRODUCTION

In 1905, in his *Handbook to the Natural History of Carmarthenshire*, T.W. Barker wrote -

> *It might well be expected that Carmarthenshire, with its ... fine rivers and beautiful woodlands, its extensive sea front and its lofty hills ... would afford a rich field for naturalists, and yet there is hardly a county in England or Wales where the study of Natural History has received less attention than in this.*

Some ninety years on, it could be argued that the same is still true. Unlike neighbouring Pembrokeshire, with its spectacular coastline and islands, there has been little tradition of visiting naturalists to record our wildlife. A colleague once described Carmarthenshire as 'the gateway to Pembrokeshire' - a place to be passed through on the way to somewhere else. Both Ceredigion and Glamorgan are lucky to have colleges of the University of Wales, at Aberystwyth and Swansea, which inevitably bring biologists to an area who will then go on to study the local wildlife. Carmarthenshire, for the most part sparsely populated by its indigenous Welsh, and Welsh speaking, population, has had to rely on a small number of dedicated local naturalists to record its wildlife. Even in popular groups such as birds, observers were few, though the newly opened Wildfowl and Wetland Trust centre at Penclacwydd, Llanelli, has brought local wildlife to many people. Yet few could argue that Carmarthenshire, from the rolling sand dunes of Laugharne Burrows to the heights of Bannau Sir Gaer, or from the flower rich meadows of the Coalfield to the hanging oak woodlands of the upper Tywi valley, provides an unrivalled opportunity for the naturalist who wishes to break new ground.

There has been even less interest in mammals than in other groups, mainly because their mostly small size and predominantly nocturnal habits make them so hard to observe. Even Barker, in his *Handbook* of 110 pages, devoted less than a page to mammals. It is illuminating to contrast his records with what we know of mammals in the county today. He regarded the pine marten as extinct, the common dormouse as local. He noted only two species of bats, the pipistrelle and the long-eared bat, the brown and grey long-eared bats not having been formally 'split' at that time. On the other hand, rabbits were abundant, the polecat and otter were still common, (even in the neighbourhood of Carmarthen) and the 'squirrel' - red of course - was dismissed almost without comment.

Yet, during this century there have been great changes in our mammal fauna. We have seen the decline and recovery of the otter, and realised that pine martens still occasionally occur. There has also been the discovery of a sizeable population of dormice in the south east of the county. On the debit side, the populations of all the nine species of bats recorded in the county have been greatly reduced, rabbits have been decimated by myxomatosis and the grey squirrel has triumphed over the red. There have also been new additions to Carmarthenshire's mammal list, with the discovery of yellow-necked mice and the recent arrival of muntjac deer and the perhaps less-well received mink.

There have also been dramatic changes in the Carmarthenshire landscape, to the point where Barker might scarcely recognise parts of the county today. Paramount amongst these have been the loss of unimproved flower rich pastures under the relentless onslaught of modern farming. Overgrazing by sheep has affected the uplands, whilst the change from a mixed farming system to one based on intensive sheep rearing, silage cutting and dairy production has changed the landscape, particularly in the west. It remains to be seen whether new subsidies for more environmentally friendly farming methods will help to arrest the decline. There have been gains however. The extensive conifer forests of Brechfa and Crychan have provided new habitat, although they have been implicated in the acidification of upland rivers. Often maligned as 'dead', generally by people who have never visited them, these forests have provided a valuable refuge for the red squirrel and possibly the pine marten. They also provide some semi natural vegetation in open areas such as rides, which are often the only areas of such vegetation in a sea of intensively farmed grassland.

The importance of Carmarthenshire's mammal fauna is underlined by the fact that virtually all the mammal species for which biodiversity action plans have been prepared - greater horseshoe bat, pipistrelle, brown hare, water vole, common dormouse, otter and pine marten - occur in the county. The only exception is the mouse-eared bat, now extinct in the UK (HMSO, 1996). The time therefore seems right to attempt some statement on the mammals of Carmarthenshire.

This work originally began as a short report intended as a supplement to the annual bird report. Very rapidly, as such projects seem to do, it grew as the amount of information, admittedly widely scattered, became apparent. Clearly, there was sufficient material to form a book, similar to the more frequently produced county floras or avifaunas. I therefore began collecting information in late 1994, which has grown to become this work.

This is not, however, the end of the story. Readers will notice that equivocations such as 'perhaps' and 'probably' pepper the species accounts. There is still much to discover about our mammals. What is the true distribution of the water vole, the harvest mouse or the yellow-necked mouse? Do additional species of bat, perhaps Leisler's bat or serotine, await discovery? Above all, does that most elusive of mammals, the pine marten, still have a self-sustaining population in the county. If *Mammals in Carmarthenshire* has a purpose, it is to record the status of mammals in the late 20th century for future readers, and to encourage an appreciation of our wildlife heritage by those of today. If more interest is taken in our mammals in future, that purpose will have been achieved. With luck, we can ensure that there are otters, dormice and bats for future generations to enjoy.

CHAPTER 2:

MAMMAL RECORDING IN CARMARTHENSHIRE - FROM ANCIENT TO MODERN

Arthur summoned to him Gwyn son of Nudd and asked him whether he knew aught of Twrch Trwyth ... Thereupon all the huntsmen went to hunt the pigs as far as Dyffryn Llychwr ... And at the clamour that was raised ... Twrch Trwyth came up ... Then he was beset by men and dogs. With might and main he went to Mynydd Amanw ... and ... then went to Dyffryn Amanw. (From The Mabinogion, trans. Jones and Jones)

This passage is one of the earliest references to mammals in what was to become the county of Carmarthenshire. It comes from *Culhwch and Olwen*, one of eleven tales that make up *The Mabinogion*. Although the stories were first written down in the 14th and 15th century, they date from much earlier. *Culhwch and Olwen* was probably first written down in the 10th century, although the material within it may be even older. One of the earliest Arthurian legends in Welsh, it tells the story of a group of magical wild boars, led by the fearsome Twrch Trwyth, who Arthur hunts across Wales. During the chase, they cross Dyffryn Llychwr, or Llwchwr, now the Loughor valley between Pontardulais and Ammanford. Mynydd Amanw is now thought by scholars to refer to the area on Mynydd Du between Tair Carn Isaf and Foel Fawr, whilst Dyffryn Amanw, or the valley of the Aman, lies between modern day Ammanford and Brynaman.

Information regarding the mammal fauna in prehistoric times comes from the cave excavations at Coygan (see Chapter 5). Material on mammals from the early Middle Ages until modern times is disappointingly rare, most coming only in passing references. This is frustrating for the modern naturalist, because the countryside would have had natural habitats in abundance at that time, many of which are now confined to isolated fragments in nature reserves and Sites of Special Scientific Interest (SSSIs). Some species formerly present in Carmarthenshire are now found only in the larger forests of continental Europe. However, there are references that occasionally lift a tantalising curtain on the mammal fauna of those times.

Hywel ap Cadell ap Rhodri became ruler of the kingdom of Seisyllwg in about 900 AD, an area which today broadly covers Carmarthenshire east of Carmarthen together with Gower and Ceredigion. Over the next fifty years he was to expand his power over most of Wales, and was given the epithet 'Hywel Dda' (Hywel the Good) by later generations. In fact it would be wrong to see him as an early Middle Ages Gandhi figure. Hywel was as ruthless as any other king of his time; he may even have ordered the murder of his brother-in-law. However, his connection with mammals in the county comes in the reason he earned his name - his laws - which were supposedly first promulgated in west Carmarthenshire in 928. According to the story, Hywel summoned six men from every cantref to a parliament at Whitland, where a monument to Hywel now stands, to discuss and formulate the laws. Although some historians have described this version of events as fanciful (Williams, 1985), it is considered likely by others that at least some kind of meeting took place in Whitland, attended by men of substance, to discuss the law (Davies, 1990). The laws themselves have another connection with Carmarthenshire. They survive as manuscripts from the 13th century, written some 300 years after they were first formulated. Although some material is considered original, they appear to have been added to on many occasions. A strong candidate for organising their compilation is Rhys ap Gruffudd - The Lord Rhys - who ruled the kingdom of Deheubarth from Dinefwr Castle during the latter half of the 12th century.

The laws themselves deal with a wide range of crimes and issues, such as murder, theft, property law and, incidentally, a section on 'The Law of Women', which was quite enlightened by mediaeval standards. The interest for the mammalogist comes in two passages; the rights of the King's Chief Huntsman, which describes when various game may be hunted, and the law relating to 'the value of wild and tame'. This give rules regarding the hunting of deer and boar, and also the value of various species. This passage includes the widely quoted value of beaver pelts as 120 pence, compared to that of the wolf, otter or fox at eight pence (Jenkins, 1990).

Other important sources from the early Middle Ages are *The Journey Through Wales* and *The Description of Wales* written by Gerald de Barri. Gerald was born at Manorbier Castle in about 1146. He became a scholar who studied and lectured at the University in Paris, before returning to Wales to accompany Archbishop Baldwin in a journey around Wales recruiting soldiers for the crusade. It was on this tour that he collected material for his book, which is still in print today (Thorpe, 1978). In it there are several references to mammals in Carmarthenshire, most notably a long description of beavers on the Teifi, apparently written partly from his own observations. Gerald's work for the Welsh church earned him the title 'Gerald of Wales', often rendered in Latin as Giraldus Cambrensis. He died in 1223 and was buried in the cathedral church in St. Davids.

Laws relating to hunting from the Middle Ages can give a useful, if limited glimpse at the fauna of that time. The Forest of Glyncothi lay in the north of the county between Brechfa and Lampeter. 'Forest' is used in its original sense, meaning a hunting preserve rather than a large wood, although many probably had a parkland appearance. Such forests had laws associated with them, which were administered by forest courts. George Owen of Henllys, Pembrokeshire, writing in the early 17th century, described how the laws divided animals into beasts of the forest and beasts of the chase. The former included red deer, boar and hares, whilst the latter included roe and fallow deer, the marten and fox. All these would have probably been common in the Forest. In a *cywydd*[1]

[1]a style of Welsh poetry consisting of rhyming couplets, with each line consisting of seven syllables

written by Lewis Glyn Cothi to Rhys ap Dafydd, he describes the forest of his youth in the early 15th century abounding with roe and red deer (Rees, 1995).

On estates such as Dynefor, hunt records continue to be our only source of information until the middle of the 19th century. Of particular note are the accounts, diaries and notebooks of the Cawdor collection, now held in the County Record Office in Carmarthen. These relate to estates in Golden Grove and Newcastle Emlyn and date from the 14th century. Many are now extremely fragile. Interpreting the information within them presents some difficulty, for many reports are anecdotal, referring to animals appearing on the dinner table or the results of a days shooting. They do, however, give a flavour of how common some game animals were then, such as brown hare, when compared to today.

The rise in the interest in taxidermy during the Victorian period provides another glimpse of the mammals in the county in past years. For ninety years, the Jeffreys of King Street, Carmarthen were one of the town's most notable families. Thomas Jeffreys, an Englishman, moved from Bristol to the prosperous and growing town of Carmarthen to continue in his business as a wood carver and gilder. However, after a visit to the Great Exhibition of 1851, Jeffreys started experimenting in taxidermy, which soon developed into a major part of his work. For example, in 1863 he preserved a seal, probably a grey seal, for a Mr Scott of Llansteffan. The Carmarthen Journal reported that *this was most effectively done ... and reflects great credit on his skill* (Dale-Jones, 1993).

After his death in 1870, his son Thomas Richard Jeffreys took over the firm. In the late 1800s the Victorian passion for taxidermy was at its height and this aspect of the firm's work became dominant. Many birds were brought in, including apparently such rarities as bitterns, spoonbills, Caspian plover and slender-billed curlew. T.R. Jeffreys assisted Barker with his *Handbook,* but sadly the records of the mammals brought to Jeffreys have, with a few exceptions, not survived. There can be no doubt that they would have formed an invaluable source of information. For instance one of the few to be documented were two pine martens shot in 1880. In a picture of T.R. Jeffreys outside his shop in 1895 a fox and otter are visible in the window. One can only speculate on what other records, of horseshoe bats, dormice, polecats or even wildcats have been lost, perhaps rotting in some attic or thrown on a burning pile of rubbish?

George and Walter Jeffreys carried on the family tradition during the early part of this century, with an interlude during the First World War. During this period such animals as the red squirrel were brought to them for preservation. However, tastes eventually changed and the business was wound up in 1936 when both brothers retired. Little remains of the many mammals that no doubt passed through the hands of this family which, perhaps unwittingly, made such a contribution to the study of natural history in the county.

As already mentioned, the Jeffreys provided much information for Thomas Barker, a naturalist who lived in Carmarthenshire around the turn of the century. Barker came from an old Carmarthenshire family, which had been in the county since 1767. He attended the public school at Harrow before returning to the county to work for his father's law practice. He lived most of his life in the Carmarthen area, particularly at his house at 'Oaklands' which still stands to the south of the town. In 1905 Barker published his *Handbook to the Natural History of Carmarthenshire*, an invaluable source of reference to all those interested in the flora and fauna of the county, and still the only book of its kind on the wildlife of Carmarthenshire. Unfortunately for the student of mammals in the county, Barker's comments on this group are disappointing, constituting less than a page. However, his contribution in other areas of Carmarthenshire natural history, particularly botany, entomology and ornithology, was immense. His early death at 51, two years before the First World War, deprived the county of one of its most notable naturalists (Carter, 1957).

Just managing to overlap with Barker was another natural historian, John Brunker. He was a keen all-round naturalist, who moved to the area in the late 19th century when his father settled from Dorset. Brunker became headmaster of the school at Llanegwad and later at the Pibwrlwyd Farm Institute near Carmarthen. From his correspondence he was apparently active in recording wildlife between 1910 and 1960. His notes and letters, which are now held in the archives of the Carmarthenshire Record Office, are a mine of information. For instance red squirrel was still common at that time, and Brunker also gives an unlocalized record for common dormouse in the Llanegwad area, where little, if any, suitable habitat for this species remains. He died in 1972 and is buried in an unmarked grave in Llanegwad churchyard (Morgan, 1982).

At the time that Brunker was active in east Carmarthenshire, the wildlife trusts in Wales began publishing *Nature in Wales*, a forum for observations on all aspects of wildlife in the Principality. One frequent contributor was Colin Matheson, Keeper of Zoology at the National Museum of Wales in Cardiff, who had earlier published his *Changes in the Fauna of Wales Within Historic Times*. Matheson went on to publish several articles on mammals in *Nature in Wales*. Mention should also be made of a large amount of correspondence that he accumulated over the years, now in the files of the National Museum. These letters are a valuable source of anecdotes on mammals from throughout Wales. Sadly, *Nature in Wales* did not survive, ceasing publication in 1987.

With the demise of *Nature in Wales*, there was a serious gap in the literature relating to the wildlife of Carmarthenshire, as well as Wales as a whole. The newsletter of the Wildlife Trust, like other county trusts, eschews scientific recording in favour of a lighter, more campaigning, style. However, Carmarthenshire is fortunate in having a dedicated number of naturalists who, in the early 1970s, formed the Llanelli Naturalists. The prime movers in this were Noel and Georgina Tallowin who had moved to the county in 1965 (Pryce, 1994). In a meeting with other active naturalists in the area in 1972, it was proposed that a local branch of the West Wales Naturalists Trust (latterly the Dyfed Wildlife Trust, and now the Wildlife Trust: West Wales) be set up in south east Carmarthenshire. However, disagreement escalated largely because of a clash of personalities and the suspicion that WWNT, with its focus on Pembrokeshire and the islands and financial constraints, had largely ignored Carmarthenshire. As a result, they formally constituted an independent organisation, the Llanelli Naturalists, holding its first annual general meeting in 1973.

A major part of the activities of the Llanelli Naturalists has been the production of a newsletter. Originally this began as a single typed sheet of recent wildlife sightings. Over the years, it has grown into a biannual production that has become a major forum for the publication of erudite articles on all aspects of Carmarthenshire wildlife. Very recently a new publication has begun - *The Bulletin of the Llanelli Naturalists*. This was initiated because of the weight of articles received for the newsletter, and will be the home of the longer articles, with the newsletter reverting to its original purpose. Articles on mammals have included subjects such as bats in the county, the Badger Group and dormice in Carmarthenshire. The author hopes to continue this tradition with an annual mammal report, although there has been a brief hiatus, caused largely by the need to concentrate on research for this volume!

In recent years, many species have been the subject of research by interested organisations or individuals. The most complete have been three reports on the status of the otter in Wales by the Vincent Wildlife Trust. Other species that have been the subject of intensive interest are the bats, particularly the horseshoe species, pine marten, polecat and common dormouse. Many of the reports produced by these studies have been used in the compilation of *Mammals in Carmarthenshire*, and appear in the reference list.

Finally, systematic recording of all mammal species dates only from the mid 1980s, when Shelagh Seery started to archive mammal sightings in the county. Most of the pre 1990 records in this volume are the result of her dedicated efforts. Shelagh, and subsequently Richard Rees, maintained the database until 1993 when Neil Mathew and the author took over the reins as joint mammal recorders for the county.

CHAPTER 3:

THE COUNTY OF CARMARTHENSHIRE AND ITS HABITATS

Carmarthenshire (vc 44), at 900 square miles or 2330 sq. km, is the largest of the Watsonian vice-counties in Wales. Formerly merged with Pembrokeshire and Cardiganshire (Ceredigion) to form Dyfed, the modern county, resurrected as an administrative unit of local government in 1996, coincides almost exactly with this 'old' county. Situated in the south west of Britain, the climate is oceanic in nature. Winters are often wet, and prolonged frost and snow are rare, except on the highest hills whilst summers are warm, but with occasional rain.

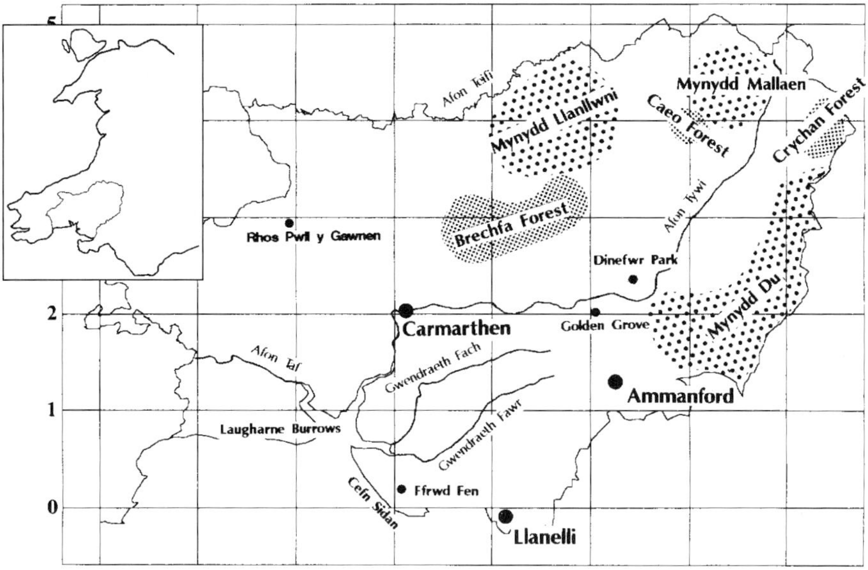

The geology of much of north and north east Carmarthenshire is dominated by rocks of Silurian and Ordovician age. Here the countryside has a rugged, upland character. Old Red Sandstone underlies much of the centre of the county, from Bannau Sir Gaer in the east to the cliffs in the far south-west, creating a plateau incised by steep, often well wooded, valleys. To the south, the Coalfield region of Pennant Sandstone and Coal Measures form the western edge of the South Wales Coalfield, producing a gently rolling landscape notable for its aggregation of unimproved pastures. Between the Coal Measures and the Old Red Sandstone is a narrow arc of Carboniferous Limestone. This area is clearly delineated, even on non-geological maps, by the line of quarries on this ridge. Where it is still relatively untouched, woodland of great significance for wildlife remains. Finally, river alluvium deposits occur, creating a more open 'fenland' type of landscape, particularly at the Llanelli and Kidwelly 'Levels'.

It is customary in county floras to give a detailed description of various habitats, as so many plants are confined to a particular habitat, or even vegetation community. Mammals are far more catholic in their tastes - one would not expect, for instance, the mammals found in a neutral grassland to differ significantly from those in an acidic pasture. In this context, only a brief description is given of the various habitats in the county to give a 'pen picture' for those unfamiliar with this most underrated part of Wales.

GRASSLAND

Unimproved species rich grassland is one of the most important of all Carmarthenshire habitats. It is also one of the most threatened, by agricultural improvement, commercial development and open cast mineral extraction.

Before the clearing of the original wildwood, grassland habitats would have been confined to clearings in the forest, kept open by the grazing of the numerous large herbivores present at that time. However, following the beginning of woodland clearance, which began in the Bronze Age, rough grassland and heath would have been an increasingly familiar part of the landscape. By the middle of the last century this habitat would have covered huge swathes of countryside quite unimaginable to the modern naturalist. For instance, Jones (1987) records a description of the Trelech, Llanboidy, Abernant and Meidrim area north west Carmarthenshire as having 500 acres of moor and 10 000 acres of 'furze and heath'. Such areas were often known by the Welsh term 'rhos' (pl. 'rhosydd'), a word which is still familiar in many place names throughout the county today.

However, the accelerating process of agricultural improvement gathered pace shortly after the Second World War, when memories of near starvation were still fresh. Reseeding and application of fertilizer has destroyed most of this resource, especially in the west of Carmarthenshire. Here, much of the grassland now consists of silage fields and heavily grazed sheep and dairy pasture, dominated by only a few high yield grass varieties. The few areas of unimproved grassland that have survived in western Carmarthenshire usually owe their existence to accidents of history. Many, for instance are areas of common land where no one owner has been able to impose change on the landscape. Most are now designated as (SSSI). Standing amongst the heather on Mynydd Ystyfflau Carn, a few miles to the north of Nantgaredig, or Rhos Pwll y Gawnen near Trelech, the modern naturalist can get a glimpse of what much of Carmarthenshire must have looked like at one time

In the east much more unimproved pasture remains. Agriculture has always been much less intense in this area. Ironically, many pastures in the Llanelli area owe their continued existence to the coal industry which until recently flourished in the Gwendraeth and Aman Valleys. Many miners earned part of their income from smallholdings and lacked the financial resources to support large scale drainage and reseeding. A few were also used as low grade grazing for pit ponies. The more rugged terrain of Dinefwr, in the north east of Carmarthenshire, has also resisted improvement. Today the distribution map for species associated with unimproved pasture, such as the delicate umbellifer whorled caraway, delineates these areas quite clearly.

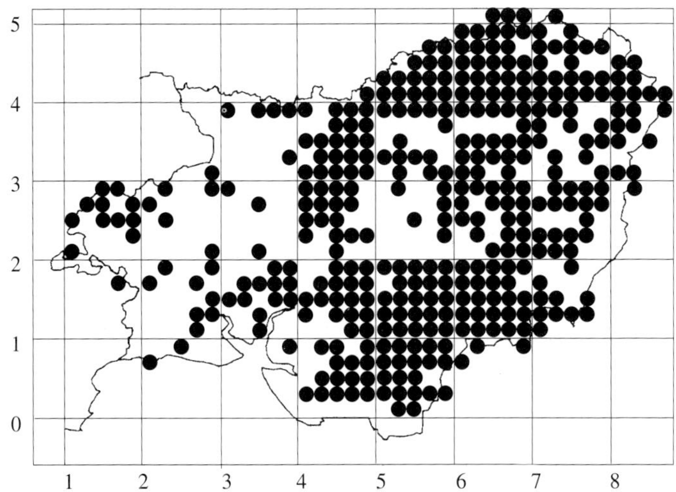

Distribution of Whorled Caraway in Carmarthenshire.

In recent years, a new threat has emerged to the remaining pastures on the edge of the coal field. Open cast coal mining has bulldozed its way across the landscape, replacing a rich tapestry of pastures and hedgerows with uniform impoverished grassland and regular, stunted hedgerows. Looking down into the Aman and Loughor Valleys from atop Mynydd y Betws, areas of 'restored' open cast are clearly visible, scarring the countryside. As such sites join up, large swathes of a historic landscape are lost forever.

Flower-rich pastures remain one of the jewels of Carmarthenshire wildlife. Amongst the mammals which use them are brown hare, most of the insectivores, dormice (which take advantage of the species rich hedgerows associated with them), mustelids and bats. These, together with numerous plants, insects and nesting birds make such places a priority for wildlife conservation in Carmarthenshire.

WOODLAND

Broadleaved Woodland

Carmarthenshire is still a comparatively well wooded county. Most of the ancient woodland which still exists clings to inaccessible slopes which have escaped the plough, especially in the west where the landscape is deeply dissected by wooded valleys, such as the Duad, Cywyn, Taf and Cych. Even here they have often been replanted with non-native tree species. In the east the pattern is repeated, culminating in the outstanding oak woods of the upper Tywi valley, including the RSPB reserve at Gwenffrwd/Dinas and the National Nature Reserve of Allt Rhyd y Groes.

Clearance of the original wild wood gathered pace in the Iron Age, when tools became available capable of making an impact on the vast tracts of forests. The Roman invasion marked the first attempt at systematic clearance, for repeated ambushes along forested roads demanded clearance alongside major routes to facilitate better defence. The slow increase in population that followed the departure of the Romans also encouraged clearance on the gently sloping valley sides as well as valley bottoms, which were used as arable land (James, 1991).

The Normans made several attempts to subdue Carmarthenshire after 1066, but Welsh resistance continued for a further two centuries. This was possible because of the large tracts of forest which could be used for ambush and guerilla warfare. One of the most notable was the primaeval forest that spread from the middle reaches of the Afon Cothi across to the Teifi, known as the Forest of Glyncothi (Rees, 1995). This played a central role in the defence of the kingdom of Deheubarth, which covered much of what was to become Carmarthenshire. Another hunting forest - the Forest of Glynne or Glynne Eiskin - was established in the Gwendraeth valley area by the Normans, who founded lordships or 'englishries' in south Wales following 1066. Gerald of Wales refers to this forest which existed in the early 12th century during the reign of Henry I, as being under the control of Maurice of London, and describes it as being well stocked with game and especially deer (Thorpe, 1978).

During the 13th century, the campaign initiated by Edward I and subsequently carried forward by later English kings, involved the systematic clearance of woodland. They intended this strategy to deny its use to the Welsh as a refuge from which to launch guerrilla attacks; a sort of mediaeval Agent Orange. For instance, in 1280 'assarting' (clearance of trees) was ordered at Laugharne, Whitland and along the strategic route from Brecon to Carmarthen (Linnard, 1982). Edward entered Wales with a well disciplined and powerful army which included carefully guarded woodmen who cleared the route, not only to provide secure passage but also to ensure timber for his castles. In a war of attrition, *the axe was his weapon as much as the lance and the bow* (Morris, 1901).

Once established, the Welsh marcher lords set up hunting forests much in the manner of those in Norman England. 'Forest' is used here in its traditional sense, meaning a hunting preserve rather than a large area of woodland. Forests such as Glyncothi and Glynne would, in any event, have been very different to the closely packed woodlands we see today. They would have included open areas of heath as well as a landscape with a more parkland appearance. The restrictive and oppressive 'forest law' governed these areas, administered by forest courts. By the 14th century, such woods were used for a large range of products. During the 16th century the Cistercian Monks at Whitland Abbey were well known for their ability to exploit woodland, including high forest and coppice with standards (Lister and Whitbread, 1988).

However, during the 15th century there was increasing pressure on woodland because of the growing population's need for agricultural land. Some areas, particularly in the west of the county were almost completely denuded of woodland. Much was also deforested because it was the haunt of thieves and highwaymen. In 1609, Gerrard Bromley, in a report to the surveyor of the possesions of the Dutchy of Lancaster, wrote that . . . *fforeste of Glynne is unknowen ... It appeares that ...* (the forest was) ... *in auntyent tyme repleyshed with goodly woode which served ...* (Kidwelly Castle). *Howe these woods have byne destroyed and wasted it doth not appeare, ffor as it should seeme it was longe since and before the memory of any nowe lyvynge* (Rees, 1953).

Some woods were protected by the ironmasters from the 16th century onwards, as charcoal was a vital resource for this industry. However, the increasing use of coke in the 18th century lead to a decline in this kind of woodland management. Wood was still valuable as a resource for pit props in the south Wales coalfield, but from the 19th century onwards, softwoods were increasingly used in mines and many native woods fell into disuse.

What little remained was devastated during the First World War and again between 1939 and 1945. The vast majority of ancient woodland had thus been lost by 1950, and much that remained was then replanted with conifers.

Much of the woodland is dominated by ash and oak, with sessile oak being most prominent in the uplands and pedunculate in the lowlands, but the hybrid often dominating. Amongst the understorey are hawthorn, blackthorn and holly, whilst willow and alder are present in the damper woodlands in the valley bottoms. However, many acid woodlands in the uplands, especially in the north and east, are heavily grazed and have little understorey. Often the ground layer is carpeted in a luxuriant growth of bilberry. Mention should also be made of the distinctive woodlands of outstanding botanical richness associated with the thin band of Carboniferous Limestone which arcs across the county from Mynydd Du down to Kidwelly.

Coniferous Woodland

Early plantings of conifers took place, from the late 1920s onwards, on the coast at Pembrey, and inland at Brechfa. At the former site, Corsican pine was the main tree used, apparently with very little early success, whilst at Brechfa a wider range of species was selected. Of particular note is the 'Forest Garden' in the Gorlech valley, which apparently was a trial area to see which trees did best in the Carmarthenshire climate. Among the trees still growing at this site are eucalypts and poplars. Many were planted by unemployed miners from the valleys, and later by Italian prisoners of war.

However, the major forests of Crychan, Glasfynydd, Caeo and Tywi were created from the 1950s onwards, and have continued to the present day. A variety of species have been used, depending on altitude, with Sitka spruce, lodgepole pine and larch species predominating in the uplands and Lawson cypress, Douglas fir and western hemlock at lower altitudes. Many forests also have a significant broadleaved component.

Conifer woods are important for several species, most notably the red squirrel. Small mammals are often abundant in newly restocked areas, whilst species of open country, such as rabbits, can be numerous on forest rides, providing food for a range of predators. Bats also occur where roosting sites are available.

UPLANDS

The uplands of Carmarthenshire are something of a disappointment to the mammalogist. Despite the presence of such relict arctic alpine plants as roseroot, dwarf willow or mossy saxifrage on the precipitous crags above Llyn y Fan Fach, no red deer, wildcats or mountain hares grace the upper slopes of Mynydd Du. The summits are cloaked in extensive grassland, dominated by mat-grass, the product of over grazing. Blanket bog would have been widespread at one time, but the combined effects of heavy grazing, burning and atmospheric pollution have largely eradicated this habitat, leaving scars of bare peat. Only small areas, notably Drysgol, still support heather and bilberry. Of some note are the cave systems which honeycomb the mountain, forming nationally important bat hibernacula.

Of the other upland areas, the plateau of Mynydd Mallaen is similar to Mynydd Du, though it receives perhaps less heavy grazing, but Mynydd Llanllwni still retains extensive heather moorland. The uplands support a range of mammal species, but there are no specialist upland mammals confined to that habitat as found elsewhere in Britain.

FRESHWATER

Carmarthenshire is drained by some of the most attractive rivers in Wales. The main rivers flow south westwards, from the high ground in the north east towards the coastal lowlands. The longest river within the county is the Tywi. It rises on the southern edge of the Cambrian

Mountains, runs through the reservoir of Llyn Brianne and on in a broad diagonal across virtually the length of the county to meet the sea south of Carmarthen. The estuary of Carmarthen Bay is also known colloquially as the 'Three Rivers' because of the two other rivers which meet the Tywi; the Taf, which drains the fertile dairy country to the west of Carmarthen, and the Gwendraeth entering from the east. The Gwendraeth is in fact two rivers draining contrasting landscapes. In the north, the Gwendraeth Fach runs through a wide agricultural valley. To the south, the Gwendraeth Fawr drains the edge of the South Wales Coalfield, with its ribbon of pit settlements, agriculturally unimproved pastures and small woodlands.

Carmarthenshire, as with so many counties, is bounded by rivers. In the south east the Loughor, which rises on Mynydd Du, forms the boundary with Glamorgan, whilst in the north the Teifi and its tributary, the Cych, form the borders with Ceredigion and Pembrokeshire.

Both the Teifi and the Tywi, together with its tributary, the Cothi, are world famous as salmon and sea trout rivers. The rivers of the Coalfield, by contrast have suffered from pollution for many years but, as the collieries have declined, so the water quality has improved. Water quality is now adjudged to be good or excellent in all major Carmarthenshire rivers (Environment Agency, pers. comm.).

In contrast to its wealth of rivers, Carmarthenshire has relatively few standing water bodies of any size. Llyn Pencarreg in the Teifi valley, Llyn y Fan Fach below Bannau Sir Gaer and Llyn Llech Owain are relatively oligotrophic or nutrient poor. Conversely, Talley Lakes, Ffrwd Fen, Machynys Ponds, Bevan's Pond, the Dinefwr Oxbows, Bishop's Pond and the only natural coastal water body, Witchett Pool, have a lowland, or nutrient rich, character. Beds of common reed (*Phragmites*) are scarce, with those at Ffrwd Fen, Witchett Pool, Llangennech and 'Ashpits Pond' near Pwll being small by British standards.

There are few man-made water bodies, though recent years have seen a spate of pond creation in the south east, for example Penrhyngwyn Pond south of Llanelli and Kymer's Canal Pond near Kidwelly. Also in this area are the now disused canals which formerly linked Burry Port with Kidwelly, together with the Swan Pool Drain and Stony Drain all which form rich habitats when appropriately managed. In the north Llyn Brianne and Usk Reservoir, which both straddle the border with Powys, are upland in nature. Lliedi Reservoir supplies Llanelli whilst Sandy Water Park lake on the outskirts of the town receives intensive amenity use. A large number of water bodies will be created as part of the Llanelli Millennium Coastal Park, including a 60 ha open water and fen complex just east of the WWT centre at Penclacwydd.

Unsurprisingly, so many freshwater sites are important as mammal habitats. Pre-eminent amongst the many species which exploit them is the otter, but they are also used by water shrew, water vole and mink. Many species of bats, most notably Daubenton's bats take advantage of the rich harvest of insect prey associated with such habitats.

COAST

Walking the large sand dune systems of the Carmarthenshire coast, it is easy to believe that you are in an unchanging landscape. In fact, nothing could be further from the truth. The dunes from Pendine to Pembrey are, in geological terms, of very recent origin, most having been formed since the Middle Ages, and much in the last fifty years.

The changing nature of the coast has been extensively reviewed by James (1991). A mediaeval map of the coast by Christopher Saxton reveals a markedly different coast to today; Laugharne/Pendine Burrows are completely absent, the coast following the line of hills that today lie behind the A4066 road. The estuaries of the Tywi, Taf and Gwendraeth are all shown as being much wider than today, particularly the latter, where the sea penetrates well inland beyond Kidwelly. Cefn Sidan is also missing with a spit at 'Caldicot Point' where Tywyn Point is today and 'Pembrey Point' where the burrows now stand. Also interesting is an island named Bachhannis off Llanelli, where Machynys Point is today.

By 1800 the dunes at Laugharne had begun to develop, together with sand hills to the south east of Pendine. However, they are bisected by the Witchett Brook, which was ultimately to form the modern day pool. The 'Three Rivers' estuary had narrowed and the skeleton of Cefn Sidan had formed. A large marsh spread southwards from Kidwelly to Pembrey, which was to be drained during the following fifty or so years. By the turn of the 20th century, the coastline had essentially assumed its modern shape (James, 1991).

All Carmarthenshire dunes would have been extensively grazed as marginal land, and were the site of the first introduction of the rabbit into the county, which served to increase grazing pressure even more. In addition, many of the dunes were extensively disturbed during the Second World War, when they were used for training in the run up to the Normandy invasion. Today much of the soft coast of the county is still under Ministry of Defence ownership as firing ranges. Most of the remainder has been either afforested or is under the care of Carmarthenshire County Council. This local authority has a large country park at Pembrey which receives intensive recreational use, and a Local Nature Reserve at Pembrey Burrows which is also open to the public. Grazing has virtually ceased with a consequent increase in sea buckthorn. As a result, the Carmarthenshire coastline is probably more stable than it has been for centuries.

Carmarthenshire has relatively little rocky coast. Cliffs spread westwards from Pendine to the Pembrokeshire border, whilst a small area of cliffs exists at Wharley Point between the Tywi and Taf estuaries.

The coast of Carmarthenshire is of great significance for wildlife. Most sections are designated as SSSIs and have been put forward for additional protection as Special Areas of Conservation under the European Union's Habitats and Species Directive.

In mammal terms, the sand dunes and salt marshes are important as large areas of unimproved land, with high small mammal populations and associated predators. The importance of the coast to the rabbit has been mentioned. It is also the refuge of brown hare, whilst water voles inhabit the associated ditches and pools. A small population of fallow deer can be found at Pendine, where the high security at the missile testing site affords them complete protection. The rocky coast is important for caves used as hibernacula by bats, whilst grey seals are seen occasionally offshore.

URBAN

Carmarthenshire is a predominantly rural county, with a population of only 169,000. The largest town is Llanelli with a population of 42,000, followed by Carmarthen (14,600) and Ammanford (8,100). Such urban areas have the usual mammals associated with the built up environment, such as fox, grey squirrel and hedgehog, but among the more unusual species recorded in recent years are otter and polecat. Bats, notably the pipistrelle, occur in houses on the outskirts of settlements where they have access to feeding habitat.

CHAPTER 4:
THE SPECIES ACCOUNTS

The following chapter gives the history and status of all mammal species occurring in the wild in Carmarthenshire, whether native or as well established introductions. Each account begins with a short summary of the biology of the species in question, before moving on to discuss its past and current status in the county. Maps and graphs use all the records, submitted to the county recorders, collected up to the end of 1996. A small number of notable records received in 1997 are also included; where this is the case, they are clearly identified in the text. Readers should be aware that *Mammals in Carmarthenshire* is not an atlas similar to county breeding bird atlases. There has, with very few exceptions, been no systematic county wide survey for mammals. Interpreting the distribution maps requires care. Most tend to be biased towards the south and east, which reflects the distribution of active mammal recorders. Only records which have a confirmed grid reference have been included on the maps. Consequently, a small number of species have no map because records did not have a definite location. Dots represent tetrads (ie squares 2x2 km), the unit commonly used for county wildlife atlases.

MAP KEY

Black dot	**records from 1990 onwards**
Dark grey dot	**records from 1970-1989**
Light grey dot	**records pre 1970**

The taxonomic order, English and scientific names follow Corbet and Harris (1991). Welsh names follow Cymdeithas Edward Llwyd (1994).

HEDGEHOG

Erinaceus europaeus

DRAENOG

A common inhabitant in both rural and urban locations.

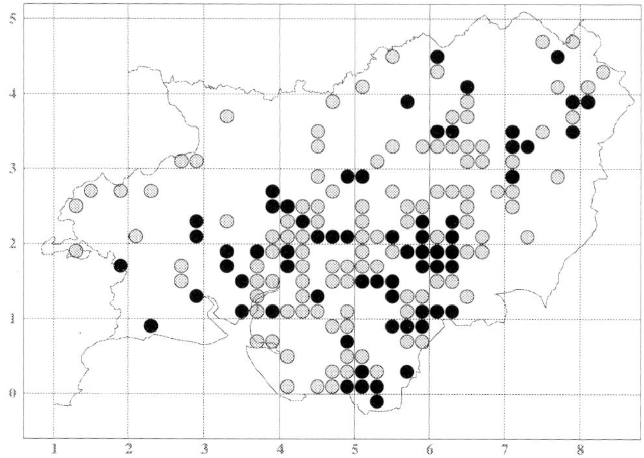

The hedgehog is a familiar sight throughout Carmarthenshire but, although common throughout Britain and Ireland, it is probably declining nationally. Loss of habitat and increasing use of pesticides are thought to be responsible, though a large number are also killed by traffic (Morris, 1993).

Hedgehogs are chiefly animals of woodland and hedgerows, wherever there is sufficient cover for them to construct their nests. They are rarer in conifer woodland and moorland, probably because there is less food available. The diet itself consists chiefly of invertebrates, and particularly slugs (Morris, 1991). Breeding takes place in June, when between four and six young are born, and occasionally a second litter may be raised if conditions are favourable. Despite occasionally falling prey to polecats and the annual carnage on our roads, failure to hibernate successfully is thought to be the single most important cause of mortality. Data suggests 65% of hedgehogs die in their first winter (Morris, 1983).

The hedgehog is widespread in Carmarthenshire. Barker (1905) regarded it as common and it has probably always been so. Though it is tempting to interpret the distribution map as indicating absence from the conifer woodlands and high ground, the pattern probably reflects observer bias. Most of the rural areas of the county are under some kind of permanent pasture, and are likely therefore to form suitable hedgehog habitat. Hedgehogs are also regularly seen in the urban areas of the south east, where they are welcomed in gardens because of their diet of slugs.

Arnold (1993) noted a spring peak in mortality, which is thought to reflect animals being killed on roads after emerging from hibernation, followed by several more peaks until September when hedgehogs re-enter hibernation. Data from the Carmarthenshire records show a similar pattern, with a rise in the number of records in April and May, followed by a larger peak in August as young hedgehogs disperse. However, it should be noted that, contrary to popular belief, hedgehogs are active throughout the year (Morris, 1995), and the Carmarthenshire data supports this.

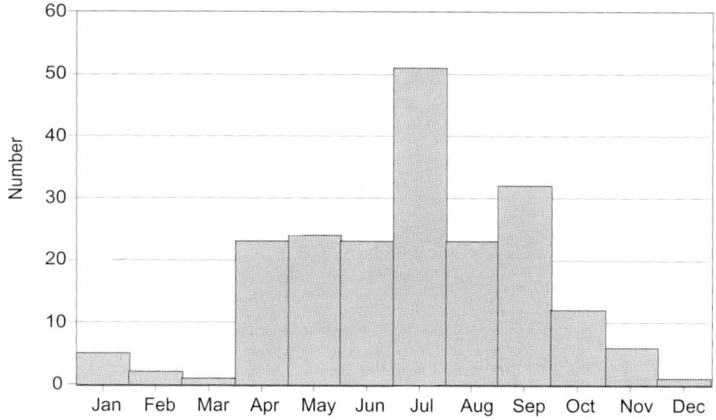

Graph 1 - Monthly Hedgehog Records in Carmarthenshire to 1996

The hedgehog is protected under Schedule 6 of the Wildlife and Countryside Act 1981.

MOLE

Talpa europaea

GWAHADDEN

Common throughout the county in grassland and woodland habitats.

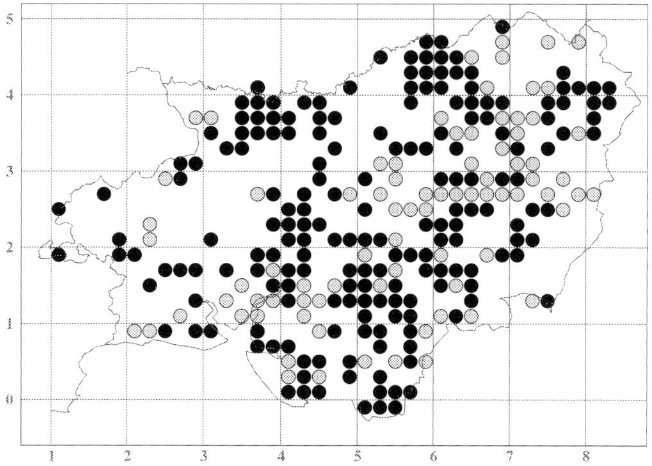

The mole is a highly adaptable species, common in woodland and pasture, and has been recorded up to 1000 m. It is found throughout Britain (Arnold 1993).

Moles are solitary mammals, spending almost all their lives below ground feeding on soil invertebrates, particularly earthworms, which they catch in their network of tunnels. Females may construct two or more nests where three or four young are born in spring, occasionally followed by a second litter. Young moles leave the nest for the first time at about 33 days and disperse after five or six weeks. Adults are highly territorial, and if a resident is removed, the territory will be swiftly reoccupied. However, during the breeding season males tunnel over extensive areas in search of females (Stone and Gorman, 1991).

Once they occupy a territory, moles are generally sedentary, but young animals are forced to disperse following weaning. This is the point at which they are most vulnerable and when predation is most severe. Moles may be taken by buzzards, stoats and cats, as well as being extensively persecuted at all times by humans.

Moles are recorded throughout Carmarthenshire. In rural areas they are regarded as a pest because of the danger their tunnels pose to livestock and machinery, and because their hills can cause contamination of silage. Much of Carmarthenshire, apart from urban areas, conifer woodlands, wetlands, mobile sand dunes and regions where the soil is insufficiently deep to allow tunnelling, is likely to provide ideal habitat.

Most mole records take the form of molehills, which are generally recorded in winter. This reflects the fact that they are more visible at this time rather than increased activity by the moles (Arnold, 1993). Molehill recording in Carmarthenshire followed a similar pattern, with a peak in January, February and March.

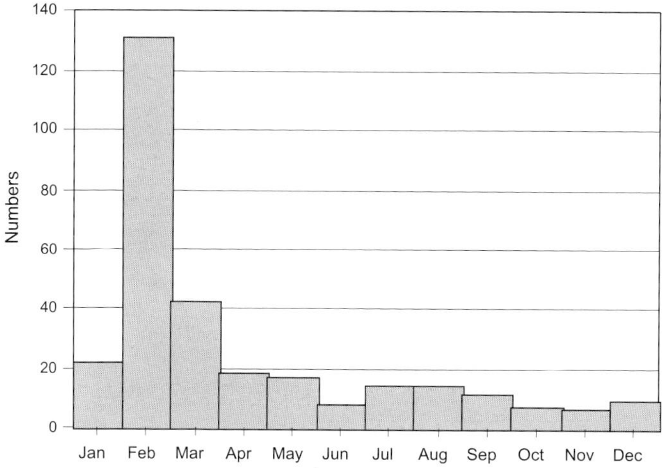

Graph 2 - Monthly Mole Records in Carmarthenshire to 1996

Moles may be vulnerable to dry weather when their prey moves to deeper, more inaccessible locations within the soil. During July 1955, the dry summer was thought to be responsible for many moles being found dead between Rhandirmwyn and Llandovery (Davis, 1955). Apart from this reference, the mole does not figure significantly in the natural history literature of Carmarthenshire. Barker (1905) regarded it as common and Thomas (1955) noted the mole as being numerous on Laugharne Burrows, presumably in damper, more stable and humus rich areas, where the sand had become consolidated and suitable for digging.

A single individual with an orange belly was found at an unnamed location in the county in 1955 (Davis, 1956). Such abnormal coat colours are thought to be more common than in other British mammal species (Stone and Gorman, 1991).

COMMON SHREW

Sorex araneus

LLYG CYFFREDIN

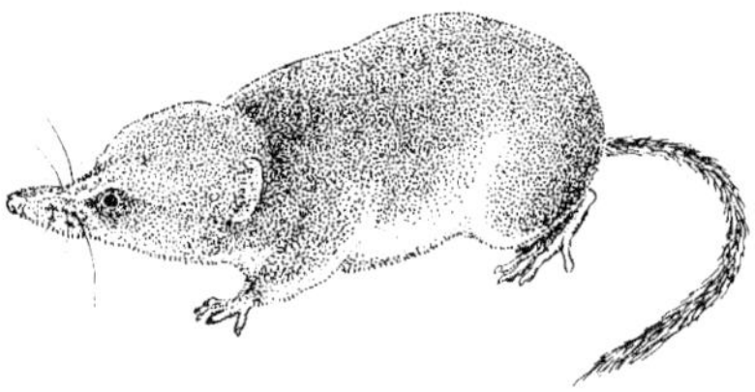

A common but under recorded species in most habitats.

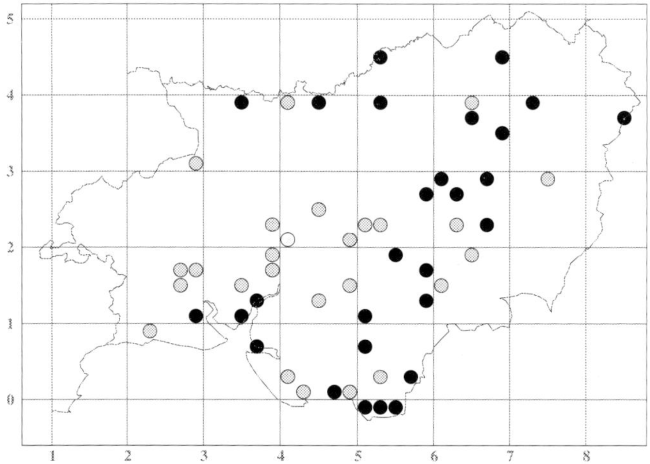

Common shrews are abundant throughout Britain especially in woods and marshes, and have been recorded up to 325 m (Arnold, 1993).

Individual animals live in territories, which the males abandon in spring to seek out females. Breeding takes place from April to September, when up to four litters of an average of six young are raised. The species has an annual life cycle, with most adults dying after breeding is finished. Consequently, common shrews seen during the winter are likely to be young animals raised the previous summer (Churchfield, 1991a). Like other shrews, common shrews are insectivorous, feeding on beetles, earthworms and woodlice. In turn they are preyed upon by a variety of animals, including owls, stoats, weasels, and cats, though they are unpalatable to the latter.

Although there are comparatively few records for this species in Carmarthenshire, common shrews are clearly widespread and are probably quite common throughout the county. The species may be scarcer in areas where its insect prey populations are lowest, such as urban areas in the south east and the heavily grazed, improved agricultural grassland, which predominates in the west. However, the species is abundant in deciduous woodland, and the wooded valleys which dissect the countryside to the west of Carmarthen no doubt provide reservoirs for this species. Common shrews have also been recorded hunting for prey amongst tideline debris on salt marshes near Llansteffan. The paucity of records reflects under recording, difficulties in observing the species and the problem of separating this species from pygmy shrew. Most common shrews are recorded in summer (Arnold, 1993), presumably because populations are at their highest at that time. The pattern is repeated in Carmarthenshire, although given the limited number of records such results should be treated with caution.

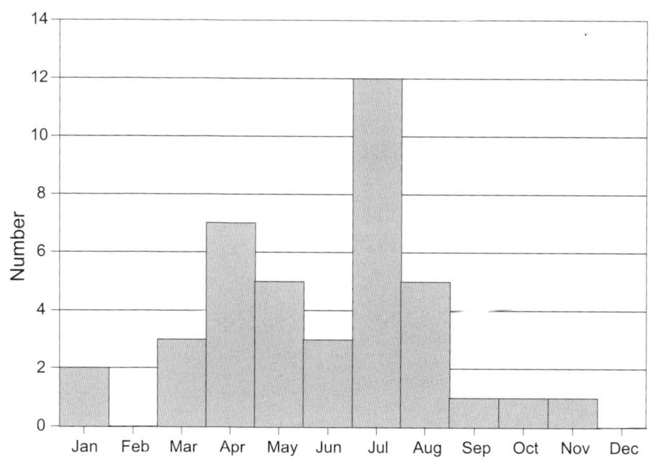

Graph 3 - Monthly Common Shrew Records in Carmarthenshire to 1996

All shrews are protected under Schedule 6 of the Wildlife and Countryside Act 1981, though this is something of a legal anomaly caused by the need to bring UK legislation into line with European law (Morris, 1993).

PYGMY SHREW

Sorex minutus

LLYG LLEIAF

A common but under recorded species, though probably less numerous than common shrew.

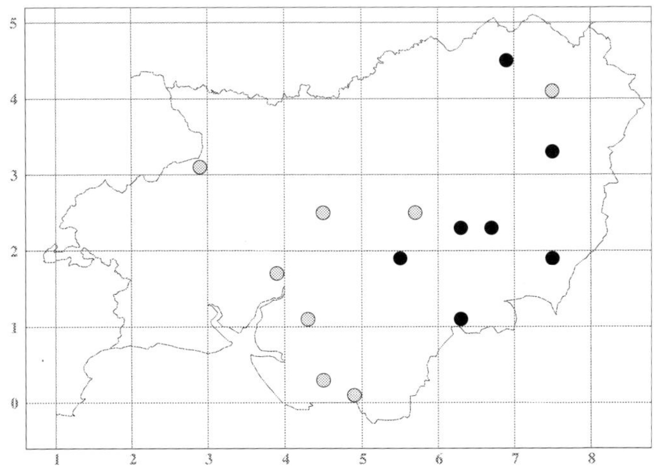

Pygmy shrews are common throughout Britain, having been recorded to 1300 m (Arnold, 1993). The species is plentiful in woods and grassland, though it is thought to be scarcer than the common shrew by a ratio of 5:1. Research in Ireland has shown that pygmy shrew is more frequent than common shrew in moorland and uplands, which may reflect ecological separation, at least in these habitats (Yalden, 1981).

Pygmy shrews are solitary and aggressive, holding individual territories, but these are abandoned in spring, especially by males as they search for mates. Females mature in March and April, and breeding takes place from April to October with a peak in June and July. Litters of four to seven are raised and, once breeding is complete, the adults die in their second autumn. Consequently, the overwintering population consists of young animals born that summer (Churchfield, 1991b). Pygmy shrews feed mainly on beetles, woodlice and flies, and in turn are preyed upon by owls, foxes and possibly mustelids.

Carmarthenshire provides abundant habitat for the pygmy shrew. It has been recorded in most 10 km squares in the county, and it is probably common throughout. Given the evidence for its preference for upland habitats it may be more common in the north and east, but there are insufficient records to discern any kind of pattern. The paucity of records reflects the difficulties of recording the species, as with common shrew.

There is a single record from *Nature in Wales* for this species in Carmarthenshire, of an individual caught in a bottle on the north shore of the Gwendraeth Estuary in August 1955 (Davis, 1956).

As with common shrew, the pygmy shrew is protected under Schedule 6 of the Wildlife and Countryside Act.

WATER SHREW

Neomys fodiens

LLYG Y DŴR

A local species, recorded in aquatic habitats.

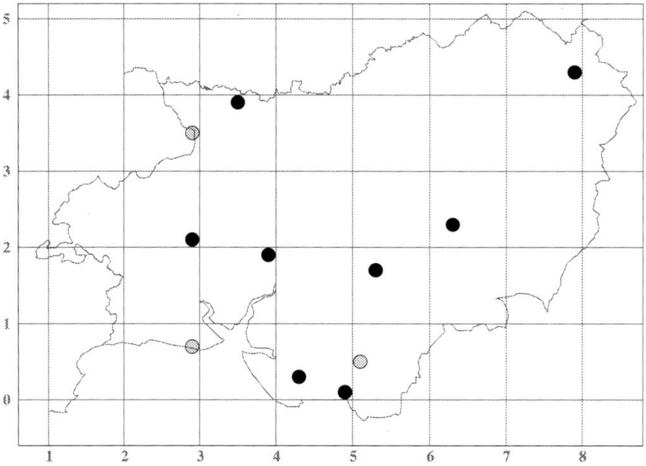

Water shrews are common, but local, on clear, fast flowing streams, ditches and ponds throughout Britain, and may be found up to 3 km from water (Arnold, 1993). Within its national range the water shrew is patchily distributed, with small pockets that are prone to local extinctions, and is much less common than the other mainland shrew species (Morris, 1993).

Individuals are solitary, feeding on land and underwater on a variety of invertebrates plus small fish and amphibians. Up to 67% of the water shrew's prey is derived from aquatic sources, especially freshwater shrimps and water lice (Churchfield, 1991c). There is some evidence that water shrews may use venom to immobilise their prey. One or two litters are raised during the summer months with an average of six young in each. Water shrews are preyed upon by birds, mammals and fish.

The water shrew is very much under recorded in Carmarthenshire but, considering the high overall quality of water in the county's rivers, it should be widespread. However, water shrews are highly influenced by human activities, and may be adversely affected by inappropriate river management, such as dredging and flood defence. The water shrew may also be an unseen casualty of river acidification in the uplands, along with more visible riverine species such as dippers (Lovegrove, Williams and Williams, 1994). Few naturalists or local people interviewed during the research for *Mammals in Carmarthenshire* had seen water shrews, suggesting the species is genuinely scarce in the county.

Although water shrews can be encountered some way from water, all Carmarthenshire records are associated with aquatic habitats, including the Afon Cych, Gwendraeth Fach, Witchett Pool and tributaries of the Tywi and Teifi. Given the number of rivers and streams in Carmarthenshire, the species could occur almost anywhere.

One individual was found with a rosy belly on Pendine Sands in about 1970.

As with other shrews, the water shrew is protected under Schedule 6 of the Wildlife and Countryside Act.

BATS

Chiroptera species

YSTLUMOD

Bats deserve a separate introduction, mainly because all the species share common problems with regard to their conservation in Carmarthenshire. A brief outline of the issues here will prevent needless repetition in the species accounts. Bats are also unusual because of the depth to which they have been studied and the unique identification problems that they present, causing many unidentified records even from experienced batworkers. As a result of the Wildlife and Countryside Act 1981, all bats and their roosts are protected by law, and the information that this has generated has revolutionised our knowledge of this previously mysterious group. However, bats, unlike other mammals, remain largely the province of the specialist.

As with much of the rest of Britain, there is virtually no historical data on bats in Carmarthenshire. Barker (1905) lists only the pipistrelle and the long-eared bat as occurring around Carmarthen. Few of the people interviewed for *Mammals in Carmarthenshire* could give any reliable information on bats. This is because bats are almost impossible to identify in the field without the specialist equipment known as a 'bat detector'. This converts electronically the ultrasound made by bats in flight into a sound audible to human ears, allowing the bat to be identified much in the way that ornithologists use bird song. However, sounds produced by bats depend on the habitat and on the activity in which the bat is engaged, and different species will produce similar sounds when flying in similar habitats. As a result, interpreting information from bat detectors requires considerable experience. Identification in the hand is no less problematical, sometimes requiring the minute observation of body parts.

During this century, two species have been 'split' by the taxonomists. The long-eared bat was discovered to be two species; brown long-eared bat, found throughout most of Britain, and grey long-eared bat, confined to the south of England. The whiskered bat was also divided into two species, with the separation of Brandt's bat in the 1970s. At the time of writing, it is likely that the pipistrelle will also yield two, yet to be named, species. Many batworkers now routinely refer to 45 kHz and 55 kHz pipistrelles (the species have the peak energy of their echolocation calls at different frequencies). Such discoveries emphasize how little we know about bats - it is as if the song thrush was discovered to be two species - and has no parallel amongst other British vertebrates since the description of marsh and willow tit in the late 19th century.

All bat species are in decline, as they come under pressure at every part of their life cycle. The loss of species rich grassland and woodland has deprived them of habitat where they can hunt their insect prey. At the same time the use of chemicals, particularly lindane, to treat pests such as woodworm in houses has indirectly killed many bats. They are very vulnerable at this time, when the breeding population of a large area may be concentrated in one building, and accidental or malicious action has decimated many colonies. Bats also breed slowly, usually raising only one young a year, so that populations are slow to recover. Finally, caves, important as sites for hibernation, have been quarried away or are being regularly disturbed by human visitors, whose body heat can arouse the bats from torpor. Such disturbance causes the bats to use valuable stored energy, which affects the animal's ability to survive the hibernation period. Consequently, and at the request of local batworkers, no individual caves have been identified in the text. The sites will be referred to as follows;

1. 'Pendine Caves'. Two sites in south west Carmarthenshire.
2. 'Crwbin Caves'. A series of caves associated with the Carboniferous limestone ridge which runs from Kidwelly to Mynydd Du.
3. 'Mynydd Du Caves'. A large number of caves and potholes in the Mynydd Du area.

Information in the text on numbers of hibernating bats at various sites should be interpreted with caution. It is well known that weather conditions can radically affect the numbers of bats at a particular location, with bats moving from site to site, or to different locations within the same site, in response to changing temperature. Because of the difficulties of observing bats wedged into crevices in caves, counts greatly underestimate the numbers of bats present. The total observed may be as little as 1% of those present.

Bats are amongst our most endangered mammals. It is no coincidence that the only mammal to become extinct in the British Isles this century was a bat; the mouse-eared bat, the last population of which had disappeared by 1980.

GREATER HORSESHOE BAT

Rhinolophus ferrumequinum

YSTLUM PEDOL MWYAF

A low density, but widely distributed resident, most evident in the winter.

The greater horseshoe bat, a large and charismatic species, is also one of our rarest and most well studied bats (e.g. Ransome, 1990). Though it may have been moderately common in southern Britain at one time, it is now restricted to parts of south and south west England and south west Wales. The total population may be as little as only 1% of the numbers found earlier this century.

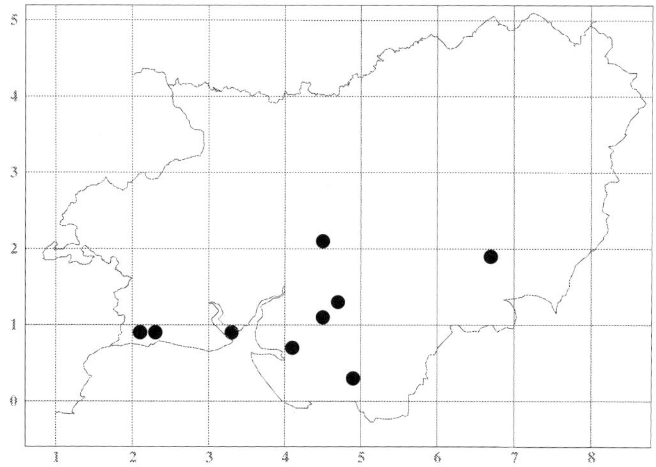

Greater horseshoe bats rely on a number of roosts in a year, which are used for different activities. Most breeding females return to the nursery site in mid June. Males may also return to the breeding site, but leave in mid July when the females give birth. In September, when breeding is finished, females visit males in their mating territories, often some distance from the breeding site. Up to eight females may visit a single male, who can hold a mating territory for up to 12 years. Once mating is completed, hibernation takes place in caves, cellars, tunnels and disused mines. In Gloucestershire, bats move up to 35 km from the breeding site (Ransome, 1991a) to find a suitable wintering site, though bats in south Wales are known to travel even farther than this.

The diet consists of dor beetles, cockchafers, moths and craneflies. Greater horseshoe bats may capture food on the ground, or hunt from a favourite perch, not unlike a flycatcher. The bats themselves are preyed upon by barn owls, tawny owls and cats, but the main cause of mortality is starvation in late cold springs, which particularly affects adult males and juvenile bats in their first year. Greater horseshoe bats have been recorded living up to 26 years in Gloucestershire.

Although the species is now limited in Carmarthenshire to low numbers of migratory individuals, it is possible that greater horseshoe bats bred in Carmarthenshire at one time. Vesey-Fitzgerald (1949) noted the species as occurring near Devil's Bridge in Ceredigion, some fifty miles north of current nursery colonies in Pembrokeshire. Perhaps some long forgotten building in Carmarthenshire was home to this species at one time? For instance, there is an unconfirmed report of greater horseshoe bats using a building near Llanboidy. The owner was apparently very annoyed by the presence of the bats and killed one with a tennis racquet (T. McOwat, pers. comm.). McOwat (1994) also recounts a story of bats, possibly of this species, breeding in a dome of an overhang in a quarry near Trevaughan. This site has now disappeared beneath vegetation and has probably collapsed. The possibility that a small nursery roost still awaits discovery in west Carmarthenshire cannot be ruled out.

However, greater horseshoe bats are still quite likely to be in the county during the summer months, on feeding forays from nursery sites across the border. Route finding and radio tracking work on bats from the Slebech nursery site in Pembrokeshire has shown that individuals can travel up to 14 km to feeding areas. This is far enough for them to potentially feed across the border in Carmarthenshire (Duverge, 1997). Small pre breeding roosts exist near to the border and some may exist within the county. Bats from this type of roost may account for two sightings in 1997 of greater horseshoe bats crossing roads in the Drefach and Saron areas (T. McOwat, pers. comm.).

Most modern records all refer to wintering individuals in various caves and mines around southern Carmarthenshire. They are conveniently split into two groups, with caves in the south west near the Pembrokeshire border supporting small numbers of bats whilst a cluster of caves to the east host single individuals.

The two Pendine Caves are regularly used by small numbers of bats. Ringing studies have shown that at least some of these are females, suggesting that they are hibernacula for groups of both females and immature animals. Data available for 1975 onwards has shown that these sites have been used almost continuously from that time. A maximum of eight bats were present in one cave in 1979 (T. McOwat, pers. comm.), whilst the largest count for the second was seven in 1994 (McOwat, 1994). Long term trends are difficult to discern, because weather strongly influences counts on a particular day. However, there is an impression of steady decline at these sites from 1976 to about 1987, with a recovery since that date (Ransome and McOwat, 1994). This is as a result of improved breeding success at the Pembrokeshire nursery sites. One of the Crwbin Caves was also host to significant numbers of this species in the past (T. McOwat, pers. comm.) with a maximum of eight in 1975 and six in 1980. Unfortunately, bat numbers have declined markedly at this site and are now confined to occasional individuals (see below).

By contrast, there is a wide range of caves, mines and other sites in the south of the county that support either single individuals or, rarely, two bats during the winter months. They include sea caves in the Pendine area, the Crwbin Caves, sea caves near Wharley Point, disused lead mines near Carmarthen, disused coal mine adits in the Llanelli area, and Laugharne, Carreg Cennen and Kidwelly castles (Lucas, 1994, 1996; McOwat, 1994). Males use these as mating territories in the autumn, and subsequently as wintering sites. It is likely that they defend these locations from other males and that the resident male actively attracts females to mate with them in the autumn. Many of these sites have only been reöccupied by bats after absences of several years, and this may reflect increased breeding success at nursery sites in Pembrokeshire.

It has been emphasised that greater horseshoe bats require a range of sites - nursery areas, hibernacula and mating roosts - to maintain a viable population in a given area. Sites in Carmarthenshire are important in supporting nursery colonies across the Pembrokeshire border. If breeding success continues to improve, we can look forward to more of these attractive bats gracing the countryside of Carmarthenshire in years to come.

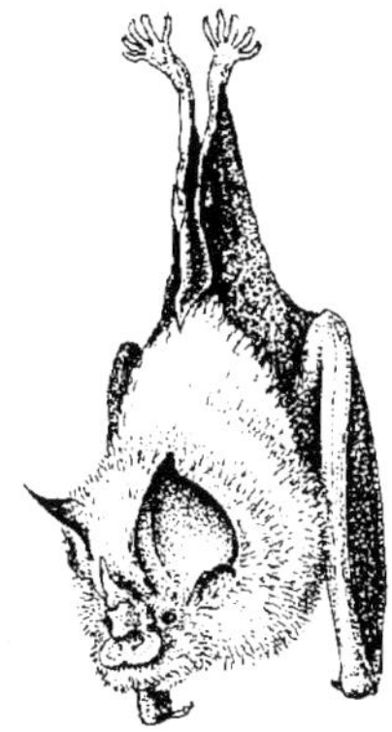

LESSER HORSESHOE BAT

Rhinolophus hipposideros

YSTLUM PEDOL LLEIAF

A scarce winter visitor to caves in the west.

The diminutive lesser horseshoe bat is found in many parts of Wales and the West Country, in sheltered valleys with extensive deciduous woodland. It has a range of preferred roosting sites including attics, chimneys, cellars, tunnels, sewers, mines and caves.

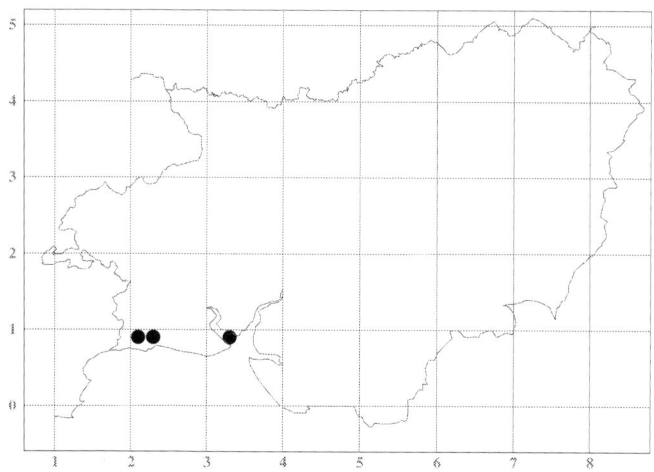

The social organization of this species is essentially similar to the greater horseshoe bat. Females in Britain do not normally breed for the first time until they are 15 months old. Juvenile bats are raised in a variety of sites, but temperature is critical and females may move if the breeding site does not offer a variety of conditions. Young are born in about mid June, though food supply and spring temperatures affect the precise date. Once the young are independent the females begin to disperse to winter sites, though hibernation starts later and ends earlier than in greater horseshoes. Small hibernacula in autumn may contain a single male together with females, suggesting a similar territorial system to greater horseshoes (Ransome, 1991b).

The diet in one study in County Clare, Ireland (Ransome, 1991b), was shown to consist mainly of nematoceran flies, with moths, caddis flies and lacewings also taken, often in dense vegetation. The species' great manoeuvrability and freedom to chew without loss of echolocation permits it to catch prey in conditions in which few other British species can operate.

The absence of the lesser horseshoe bat as a breeding species in Carmarthenshire is something of a mystery, since there are colonies close to the border at Colby Lodge Pembrokeshire, on Gower and in Breconshire. It is even possible that a small summer roost awaits discovery in the south of the county in the Pendine/Laugharne area. There is a report of this species breeding in the church at Marros in the 1970s, where a local schoolteacher would take children in to see the bats. Sadly, the church tower where the bats were rumoured to be breeding was cleared out in the 1980s and the bats have not been sseen since. It is possible that the absence of suitable wintering sites may also be preventing the return of the species.

Lesser horseshoe bats are known from only three locations during the winter months. The Pendine Caves support the bulk of the hibernating population. The maximum count for both sites is five in 1988 and 1989, although larger numbers - possibly up to 30 - have been seen swarming in the area (T. McOwat, pers. comm.). Losses of feeding habitat such as unimproved grassland may have contributed to the decline of this species at the latter site (McOwat, 1994). Sea caves at Wharley Point also support small numbers - less than 3 - on a regular basis. However, it is quite possible that many lesser horseshoe bats go unrecorded, because they hibernate in caves too small for humans, or even in places such as badger setts.

WHISKERED/BRANDT'S BAT

Myotis mystacinus/brandtii

YSTLUM BARFOG/BRANDT

The two species are inseparable unless observed at close quarters. Many records simply refer to whiskered/Brandt's bat.

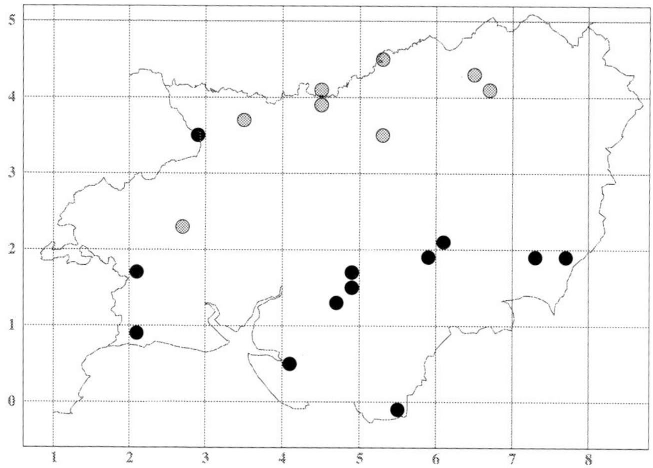

WHISKERED BAT

Myotis mystacinus

YSTLUM BARFOG

A widespread species, in buildings and caves. Status uncertain due to identification difficulties.

The whiskered bat is probably found throughout Britain. However, its distribution is uncertain due to the confusion with the very similar Brandt's bat, from which it was only separated as late as 1970. It is found in open country and woods, roosting in trees, buildings and occasionally bat boxes in summer, and caves in winter.

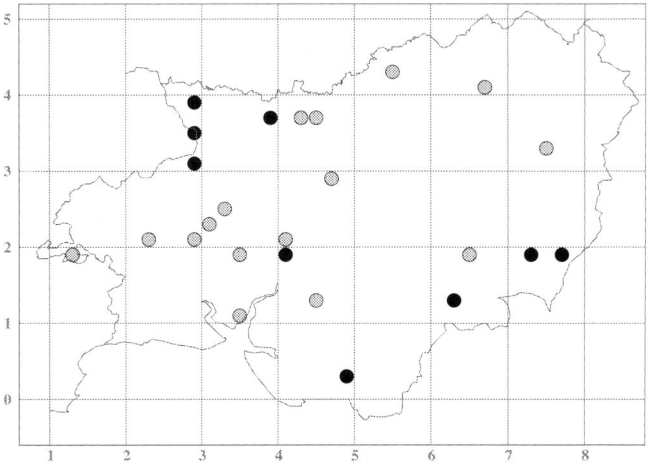

In May females form nursery colonies of between 20 and 70 individuals. The birth of the single young takes place around mid June and, once the young bats are independent, the colony begins to disperse, usually around the end of August (Schober and Grimmberger, 1989). By contrast the males lead separate and solitary lives. In winter they utilize the colder areas of caves around the entrance, with bats hanging freely or squeezed into tight crevices. Trees and houses are probably also used in winter to some extent.

Whiskered bats feed on small insects and spiders, some of which they may glean from vegetation.

The species is particularly associated with the western side of Britain (Mitchell-Jones, 1990). It is widely, if locally, distributed in Carmarthenshire, even in the west of the county where steep sided valleys cloaked in deciduous woodland offer it suitable habitat. Counts at breeding sites are few as the species is rarer than the pipistrelle, the usual bat to be found in buildings, and because of the difficulties in separating this species from Brandt's bat. In 1993, 29 whiskered/Brandt's were counted at Henllys near Cilycwm (Lucas, 1994). However, the ease with which whiskered/Brandt's bats have been recorded on guided walks around the county using a bat detector (Lucas 1994, 1995) suggests that there could be many nursery colonies undetected around the county.

Whiskered bats are encountered far more often whilst hibernating at cave sites around the county. Small numbers can be found in caves in the Pendine area and in the Crwbin caves, but the bulk of the records emanate from a handful of caves on Mynydd Du. One cave is particularly notable with, for instance, 32 whiskered/Brandt's bats present in 1994 (McOwat, 1994). This cave is quite short and difficult to enter and is consequently rarely visited by cavers, which no doubt adds to its attraction for bats. As a result, bats can often be seen clustering, and groups of eight or ten whiskered/Brandt's bats are not unusual. Another cave in the Mynydd Du complex is also an important site though, in contrast, it is heavily used by cavers. Fortunately, most bats are found around the entrance and are missed by visitors as they move to the more challenging caving farther inside. The largest count for this site is 17 whiskered/Brandt's in 1994 (McOwat, 1994). A handful of other caves in the area also support smaller numbers of wintering bats.

BRANDT'S BAT

Myotis brandtii

YSTLUM BRANDT

A widespread species, in buildings and caves. Status uncertain due to identification difficulties.

Brandt's bat was only split from whiskered bat in 1970, and therefore there is great confusion over the distribution of the two species. It is thought to be widespread throughout Britain in wooded country, utilizing caves and buildings (Stebbings, 1991).

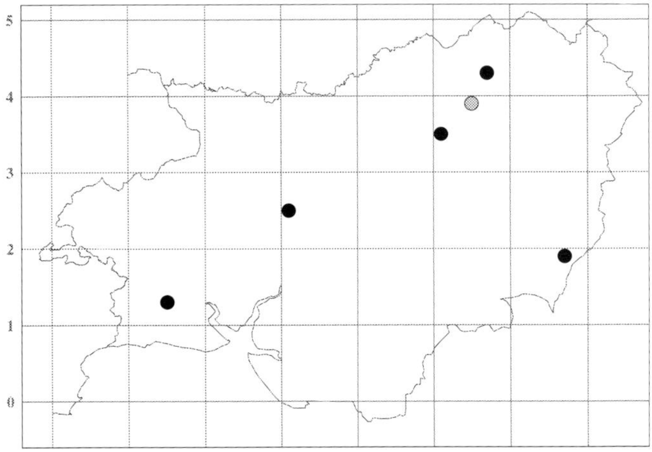

In summer, the females segregate from the males to form nursery colonies, often under the ridge tile in buildings. Births take place from mid June to mid July, of one young, which can fly by the age of three to four weeks. Adult males are solitary. In winter they occupy caves and cellars. Brandt's bats feed on small insects and spiders that are sometimes picked from foliage.

Confusion over the identification of this species, which is virtually identical to whiskered bat, continues to cloud its true status in Carmarthenshire. Readers are advised to refer to the account for whiskered bat which refers to many unidentified records. Nursery sites are few, but well distributed, suggesting that, like whiskered bat, this species may be widely, if thinly, scattered throughout the county. There is no data regarding the size of any nursery roost.

Information on wintering sites is also scanty, with many winter records referring to whiskered/Brandt's bats (see whiskered bat account for details of these). A count of 26 bats in one cave on Mynydd Du is the best for the county. The species is thought to be widespread at hibernating sites, much as whiskered bat.

NATTERER'S BAT

Myotis nattereri

YSTLUM NATTERER

Widespread, but few breeding sites known. Regular in caves in winter.

Natterer's Bat is widespread in Britain, in parkland and woodland. Both hollow trees and buildings are used in the summer months, whilst caves are the preferred hibernation site.

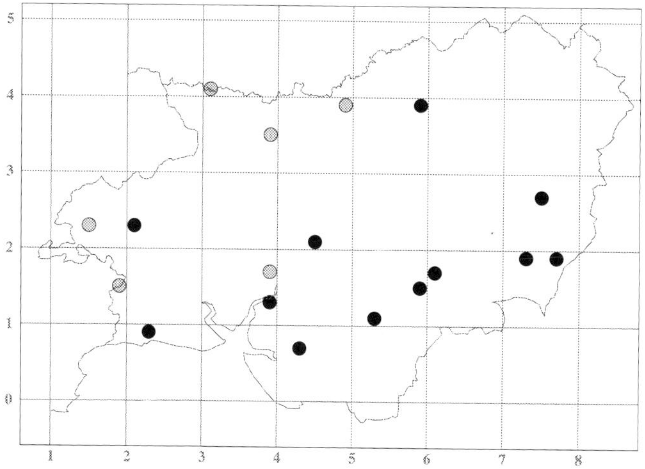

During the summer months females form nurseries of approximately 75 - 200 bats in buildings. The birth of the single young occurs from mid June through July. In winter Natterer's bat prefers cooler areas in caves and is often the most common bat at exposed sites. Most bats arrive at the hibernation site between September and December. The diet consists of moths and caddis flies, caught mostly in flight but also taken from foliage. The species association with crevice sites, in buildings, caves and walls, means that it is easily overlooked

Little is known of the status of Natterer's bat in Carmarthenshire, although the number and distribution of records suggest it is widely scattered in the county. Blackmore (1955) records this species as probably common throughout Wales, and it is unlikely that its status in Carmarthenshire will have changed significantly since then. Mitchell-Jones (1990) also noted a westerly bias in the distribution of this species, though it was less marked than in whiskered or Brandt's bat. Summer roost counts are few, the largest being 170 at Derwydd

Mansion, near Llandybie, in 1995 (Lucas, 1996). The bats use the mansion loft, but exit by flying through a hole in the ceiling into a disused room, and then out through an open window. The residents have allowed bat workers to sit in the room whilst emergence is in progress, and enjoy the experience of numbers of bats wheeling above their heads before setting out for their night's feeding. Smaller numbers have also been noted near Kidwelly, where 32 were present in a house in 1993 (Lucas, 1994). During survey work, batworkers have also recorded individuals at several locations in the south east of the county, where their bat detector signal, described as like 'the crackle of distant burning stubble' has been heard.

Like many *Myotis* species, Natterer's bat is recorded more often in winter hibernacula than at summer sites. Cave sites on Mynydd Du tend to have the greatest numbers of this species. Trends are difficult to discern, because weather and degree of disturbance can influence the number of bats seen on any one occasion. However, a total of between ten and thirty bats are regularly recorded in the various caves each winter. Small numbers - less than ten - are also found in a range of caves and mines across the county, including the Pendine and Crwbin Caves. Natterer's bats have also been found hibernating in a disused mine near Llangadog (Lucas, 1996). Two males were also found hibernating in a corrugated iron shed near Penygroes in 1993 (Lucas, 1994).

DAUBENTON'S BAT

Myotis daubentoni

YSTLUM Y DŴR

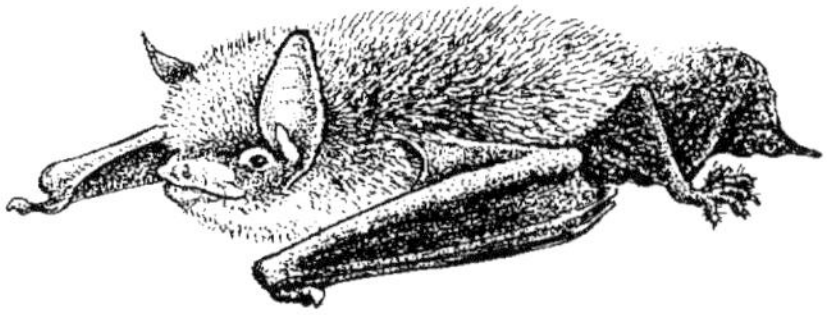

Probably widespread, near water. No breeding sites known, regular in caves in winter.

Daubenton's bat is common and widespread in England and Wales in open or wooded country, and its affinity with rivers often earns it the name, the 'water bat' (Barrett-Hamilton, 1910-11).

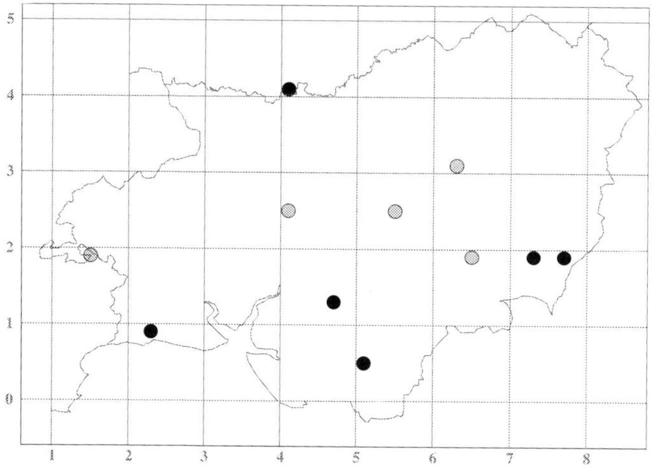

Nursery colonies of several hundreds may form in hollow trees, buildings or bridges, made up mainly of adult females. Young are born in June or July and fly at six to eight weeks. Males live alone or in small groups, in sites such as small cavities under bridges. In winter it is found in caves or scree, entering hibernacula around October (Speakman, 1991). It is a highly specialized bat, trawling the water surface for insects, such as small flies and caddis, with its large feet (Swift and Racey, 1983).

Daubenton's bat has been recorded widely throughout Carmarthenshire but, despite the number of records, there is still no confirmed breeding location for this species in the county. It can be readily observed flying a few centimetres above the surface of rivers and lakes, at such places as the Afon Tywi near Llandeilo and the mill pond at Dinefwr Park. Approximately ten were present at the last site in 1995. There are also anecdotal reports of fishermen having accidentally caught this bat whilst fly fishing on the Tywi. Daubenton's bat is rarely recorded in houses, favouring trees and bridges, making location of nursery sites much more difficult than in other species more associated with human habitation. Such roosts can be a long way from water, and in such locations hedges are important as flight lines between the roost and water. Undoubtedly, the understandable concentration by the few batworkers in the county on safeguarding house roosts has meant that this species is very under recorded.

As with related species, most records refer to winter hibernacula on Mynydd Du and other caves in the county. The species is regularly recorded in small numbers from the caves on Mynydd Du, and more rarely from other underground sites associated with limestone in the Carmel, Crwbin Carmarthen and Kidwelly areas, and also near Pendine.

Daubenton's bat remains one of Carmarthenshire's most elusive bat species. No doubt a systematic search of river valleys in the county would yield many more records of this species

NOCTULE

Nyctalus noctula

YSTLUM MAWR

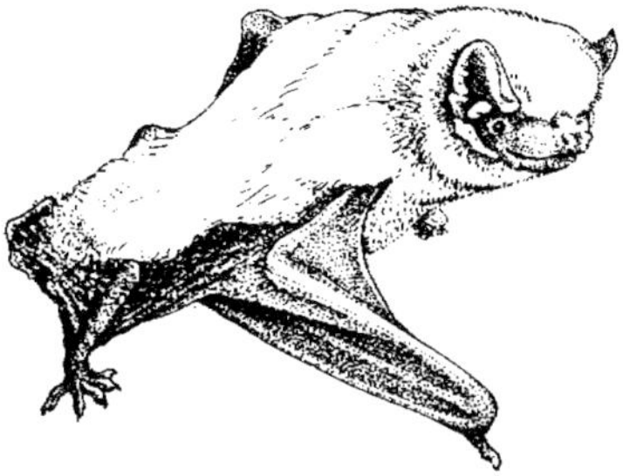

Widespread, associated with woodland.

The noctule is found throughout England and Wales, and is very much a tree bat, even using arboreal sites for hibernation. Females form nursery roosts in March and move to new sites regularly throughout the summer. The mean number of roosts found in one study was 14 (Howes, 1979). Young are generally born in the first half of June to early July, and begin to fly from early July onwards, although they are not weaned until August. The males are solitary or in small groups, and establish mating roosts in August or September. Mixed sex colonies are formed for hibernation. The main food items are house crickets, summer chafers and cockchafers, eaten in flight.

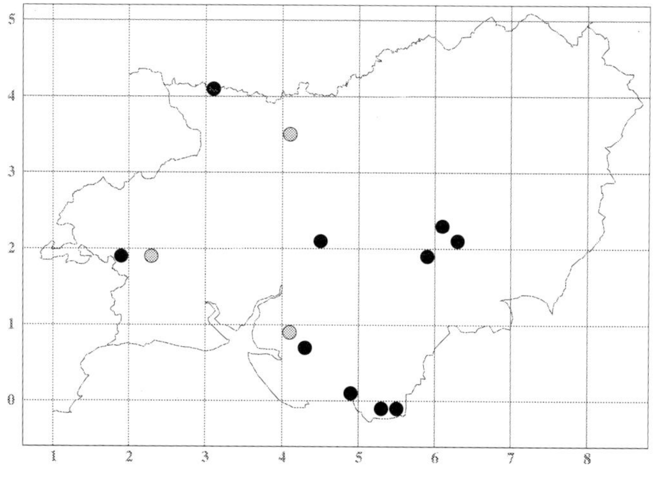

Considering the species' association with trees, it is hardly surprising that the species is widespread in Carmarthenshire with its many wooded valleys. Blackmore (1955) records the noctule as being widespread in wooded areas in Wales. This fast, high-flying species is still a familiar site in the county, especially in areas where large trees, which provide suitable breeding and hibernating sites, are situated. Mathew (1991b) recorded a colony of males in a hole in an ash tree near Derwydd in 1989, where at least 12 individuals were present. Other tree roosts have been found near Kidwelly, Whitland and Pendine, the latter of which had a pile of droppings associated with it some 40 cm deep. The roost at Whitland, in 1984, had over 100 individuals (T. McOwat pers.comm.)

Given its requirement for tree cavities, parkland areas are often the best places to search for noctules. The species has been recorded at Dinefwr Park, in the Tywi valley, where 20 were seen near Llandeilo in 1994, Talley, Abergwili, Gelli Aur, the Cothi Valley, Glanrannell and the Penclacwydd Wildfowl and Wetlands Trust Centre south east of Llanelli, where their large size and distinctive bat detector signature make them easy to identify. Noctules are vulnerable to the loss of tree roosts, for such old trees are often mistakenly regarded as dangerous or unsightly, and undoubtedly many sites have been lost in the county.

Noctules are quite unusual in house sites in Carmarthenshire, apart from one notable location near Kidwelly, where a pre breeding roost of noctules was discovered during 1993. This site has been monitored annually, with a maximum of 64 bats present in 1995. The noctules are usually present from late May through to mid June, presumably moving to alternative sites as the season progresses. The owners of the property have been particularly sympathetic in allowing batworkers access to the loft on a regular basis.

No noctule wintering sites are known in Carmarthenshire, and it is presumed that they pass the colder months in tree cavities. The species is highly migratory on the continent, and there is one record of an individual seen to fly out to sea from Pendine in the direction of Devon (T. McOwat, pers. comm.).

PIPISTRELLE

Pipistrellus pipistrellus

YSTLUM LLEIAF

Common in suburban and rural areas.

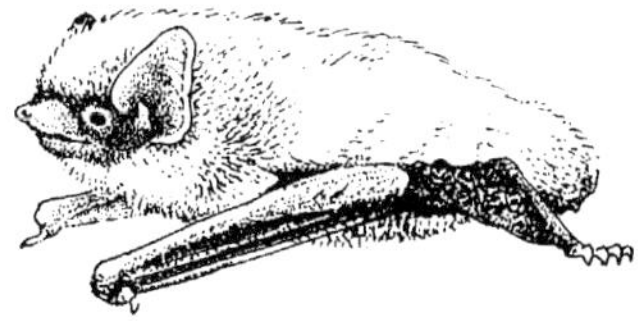

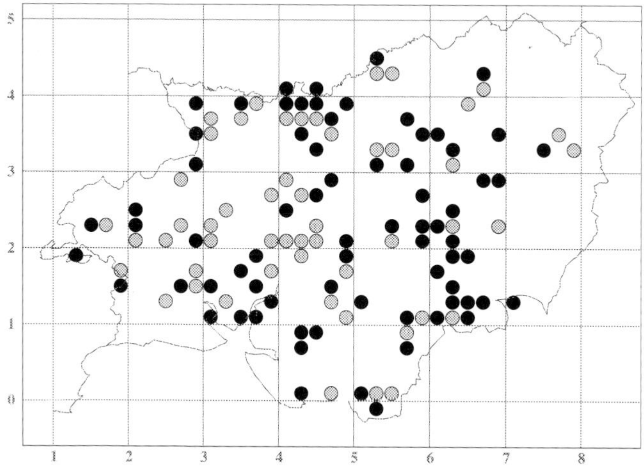

The pipistrelle is by far the commonest bat species in Carmarthenshire, probably outnumbering the combined total for all other species. Though it has been thought of as one species, it has recently been discovered that there are two 'phonotypes' (bats whose echolocation calls have different peak energies) which appear to be separate species (Barratt *et al.*, 1997). Both forms, currently referred to as 45 kHz and 55 kHz pipistrelles, are treated together since there is little information on their differentiation in Carmarthenshire, except to say that both are present.

Pipistrelles are common throughout the British Isles, feeding in farmland, woodland, suburbia, marshes and lakes. It roosts in all kinds of human habitation, from mediaeval churches to modern houses. Pipistrelles also occupy tree sites but this is rarely recorded. It is also rarely encountered in caves.

The sexes live apart in summer, the males leading a solitary life that has been little studied. Mating takes place in the autumn and males and females spend the winter together. Young are born in July and fly for the first time three weeks after birth. Feeding occurs in many habitats including farmland, woodland, suburbia and wetlands, the diet consisting of small flies, caddis, mayflies and lacewings.

Pipistrelles are found throughout the county, apart from the highest peaks, feeding wherever woodland or pasture offer insect prey. The species has been noted as roosting in a variety of structures in Carmarthenshire. Amongst the buildings used are houses of all ages, chapels, schools, a Woolworths store, the Miners Welfare Club in Ammanford, an electricity substation (both in the walls and between cooling fins), offices and the control tower at RAF Pembrey. Colonies in suburban areas frequently occupy whole streets, especially where houses, which have all been built to the same design, provide similar access points. Soffits are a favoured location, but roof spaces and the gaps between slates and underfelt are also used. Most colonies have a few tens of individuals present, but a small number of roosts have recorded many more. A building near Llangunnor currently holds the county record for a pipistrelle roost, where 523 were counted in 1995. Another exceptional roost is an unoccupied house near Brechfa, situated in splendid isolation in thick woodland where more than 450 bats were recorded in 1995 (Lucas, 1996). Other high scoring sites include the Primary School at Cwrt Henri with 499, a house near Llanddowror with 447, the Rhosnewydd estate in Tumble with 300, 300 near Crwbin, 250 in a house near Llandovery and approximately 200 bats in a house in Trap (Lucas, 1995, 1996). Such large roosts are thought to be formed by the 55 kHz phonotype. Most roosts throughout Carmarthenshire have between 20 and 60 individuals, which are thought to be formed by the 45 kHz phonotype

Pipistrelle behaviour in winter is much less well known. The species is virtually unknown from cave sites, although individuals are very occasionally recorded in the Mynydd Du caves. One notable find of a wintering site was at Newton House, Dinefwr Park, when some 200 bats were found in torpor in January 1994 behind panelling in an unused part of the house. Presumably other 'cold' buildings, such as churches, as well as walls, rock crevices and trees are also used for hibernation.

Because the pipistrelle is relatively common, it can be easy to dismiss the species as not being one of concern. However, pipistrelle numbers have fallen by some 70% since throughout Britain since 1978 (Anon., 1997). Its reliance on buildings makes it very vulnerable to persecution and unintentional killing by householders. The good work undertaken by local bat enthusiasts, to persuade people to take a more sympathetic attitude to their bats, has saved many roosts. No doubt this species will require more efforts on its behalf for some time to come.

BROWN LONG-EARED BAT

Plecotus auritus

YSTLUM HIRGLUST

Widespread, in older buildings near woodland.

This species occurs throughout Britain in lightly wooded areas, utilizing buildings and trees throughout the year, as well as caves in winter. The brown long-eared bat is the second most common species, and is often found in older buildings.

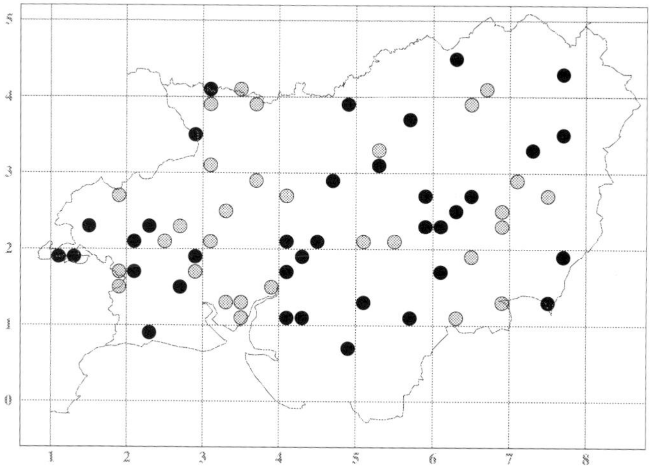

Nursery colonies of adult females, which usually number between 20 and 80 animals, form in May or June. Young are born in the first three weeks of July, or even earlier if the weather has been exceptionally dry and warm. The young bats leave the nursery roost for the first time after 30 days, and weaning is complete at about six weeks. Hibernation occurs from November to March, in trees, buildings or caves, where they prefer crevices in the cool outer regions. Brown long-eared bats are thought to undertake shorter movements than other species, and may even occupy the same roost all the year round in some areas (Howard, 1995).

Feeding takes place in woodland, where they glean prey such as moths and other insects from vegetation. The long ears and large eyes suggest that prey is located by sight and sound rather than by echolocation. Brown long-eared bats may themselves fall prey to barn and tawny owls, as well as cats.

Brown long-eared bats are widespread in Carmarthenshire, favouring older buildings such as old houses, churches and chapels as summer roosts. Systematic counts of such roosts are few, however, with 30 being recorded at a chapel near Croesyceiliog (Lucas, 1994). Most roosts are of this size, but there is an unconfirmed report of an exceptional roost in Garnant, which may have held up to 200 animals in late summer 1994 (Lucas, 1995). The species is rare at Carmarthenshire cave sites in winter, and generally occurs only during cold weather. Two bats, for example, appeared at the dungeon cave at Carreg Cennen Castle in November 1993 (Lucas, 1994).

RABBIT

Oryctolagus cuniculus

CWNINGEN

A widespread and common species of open habitats, though much reduced since the 1950s.

The rabbit is common throughout Britain and may even be increasing, though populations fluctuate according to outbreaks of myxomatosis. The species is not native, having been introduced by the Normans, but is now widely naturalised and is common in short grass or closely grazed agricultural pasture. Rabbits tend to be less common in coniferous woodland, apart from the edges and along firebreaks.

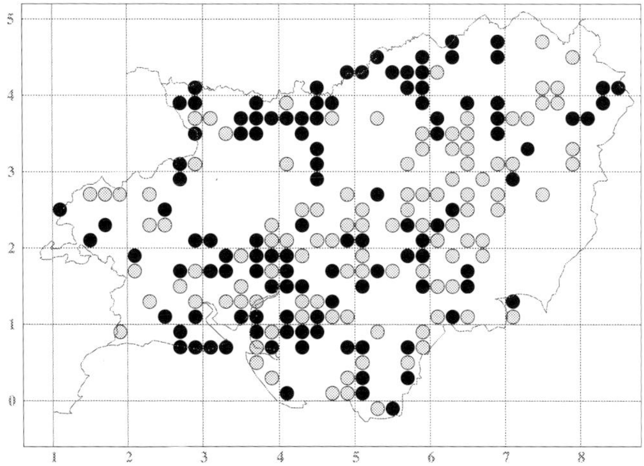

Rabbits live in stable social groups with both male and female hierarchies. Breeding takes place from January to August, when a succession of litters of three to seven young are born. The young appear at the burrow entrance at 18 days, and are weaned at 21 to 25 days. Rabbits feed on a wide range of vegetation, especially grasses, and tree bark (Cowan, 1991).

Rabbits were first introduced into Britain in the 11th century, and arrived in Wales in about 1300. Initially they were raised on islands, such as Caldy Island, which provided a secure, predator free environment (Matheson, 1941). In Carmarthenshire, which has no offshore islands, rabbit husbandry was confined to coastal sand dunes, where their presence was an additional source of protein for those inhabiting the 'englishries' of the south of the county.

Indeed, the epithet 'Burrows', which in modern usage is a synonym for 'sand dunes', suggests how wide the use of such places was for the propagation of rabbits. The coastal dunes at Pendine, Laugharne and Pembrey are all referred to as Burrows, as are many other places in south Wales and beyond. Such sand dunes, which were of marginal use as agricultural land provided ideal conditions for the construction of large warrens.

The earliest reference to rabbits in the county is a note on a warren in the parish of Kidwelly and Caldicote (a small hamlet that stood approximately where Pembrey Motor Racing Centre is today) in the 15th century (James, 1991; James, in prep.), where a grant of a warren was made in 1610. Undoubtedly rabbits were present in large unfenced warrens on Carmarthenshire sand dune systems in the mid 17th century, at about the same time that they became established in neighbouring counties. (Thompson and Worden, 1956).

Rabbits were a most valuable and jealously guarded resource. John, first Lord Ashburnham, recorded during a visit to Carmarthenshire in 1687 that *I saw the warren and ... an abundance of Rabbets there are. He* (Charles Dalton, one of Lord Ashburnham's tenants) *killed last year 400 couple and the warren will yield in two years more treble the number and be worth about £30 per annum.* (Mee, 1892).

By the beginning of the 18th century many estates had their own private warrens. Examples of such warrens remaining in place names include the 'Cwnigaer', literally 'rabbit fortress', in Dinefwr Park, and Warren Wood in the parish of Llanfihangel Aberbythych. Pillow Mounds, or artificial warrens, can also be found at Beddau'r Derwyddion, near Llandeilo (Roese, 1979), Abermarlais Park near Llansadwrn and near Llandissilio. The design of these mounds varied from the sophisticated to the crude. Generally they consisted of hummocks capped with turf, evenly spaced to prevent overgrazing, but were occasionally little more than a collection of old trees and other debris covered in earth. Tunnels would be bored into the mound and unwilling rabbits pushed in with long poles to encourage them to adopt their new homes. Such structures were raised above the water table to prevent animals drowning or dying of exposure in cold ground. Open warrens could also be found, mainly on common land (such a warren was present at Dryslwyn, for example), where they were fenced to keep the rabbits in and predators out.

However, it was not long before they escaped into the wider countryside and found the mixed agricultural landscape of Carmarthenshire much to their liking. By the early 19th century, persecution of predators such as the wildcat, polecat, pine martin, buzzard and red kite was accelerating. Not only were large areas of the county suitable rabbit habitat, but it was also denuded of predators that might have held the rabbit in check. Changes in agriculture, with the cultivation of winter fodder for livestock, also increased the food supply, and rabbits were also actively conserved as a source of winter meat. On the estates, from which they had originally escaped, numbers were also high. Data from the ten years from 1896 for the Dinefwr Estate (Carmarthenshire Archive Office ref. Dyn 275/6), give a flavour of the scale of the slaughter.

YEAR	TOTAL	HIGHEST DAILY BAG
1896	1,496	651
1897	1,328	508
1898	1,034	615
1899	759	441
1900	1,045	650
1901	1,142	705
1902	1,207	730
1903	1,164	605
1904	1,209	751

The highest bag for a single day was on the 6 September 1904, when 751 rabbits were shot. Amazingly, 418 were shot at the same location the following day. At the same time, similar bags were being taken on other estates in the county, such as Cynghordy. Whilst shooting on this scale appears senseless to modern eyes, it should be remembered that the success of a day's shooting in Edwardian times was measured largely by the bag size. Prowess with the gun was seen as an essential attribute for a gentleman and a good day's sport a tribute to the gamekeeper responsible for the estate (Tapper, 1992).

By the beginning of this century, rabbit numbers were astronomical. Carmarthenshire was noted as one of the areas to be worst affected (Condry, 1981), and Thompson and Worden (1956) regarded the rabbit population as probably amongst the highest in Britain. Itinerant gangs were employed in an attempt to control the population. However, because they would only trap to a point where the catch became uneconomic, they usually only removed the surplus produced each year. As a result, a basic breeding stock would be left to repopulate the landscape the following year. On occasion, the proportion removed by a trapper could be as little as 40% (Thompson and Worden, 1956). However, rabbits could be quite rare in the south of the county during the early part of this century, as they were heavily trapped for food during the General Strike of 1926.

By the end of the Second World War, the trade in rabbit meat was a significant part of the Carmarthenshire economy. For every 100lb of domestic meat exported from the county, such as beef and lamb, Carmarthenshire supplied 10lb of rabbit meat. More than three million rabbits were killed in Carmarthenshire, Pembrokeshire and Ceredigion annually (Thompson and Worden, 1956). The growth in the milk trade to London also assisted the trade in rabbit meat, for the special trains that transported milk to the city could also carry the product of the day's rabbiting in west Carmarthenshire. Up to 50 might be seen in a field in the Llandeilo area in the early 1950s, being particularly abundant at harvest time.

It was not until the advent of myxomatosis in 1955 that the rabbit went into decline. It is impossible to overstate the impact this disease had on the species, which is thought to have killed up to 99% of the population in Wales. Even the original strongholds of the rabbit in Carmarthenshire were devastated. Thomas (1955) noted that there were 'hundreds of thousands' of rabbits on the Laugharne Burrows in January 1955. By the end of that year the effects of myxomatosis meant there were none to be seen. In the Myddfai area, there was a large rabbit population in the early 1950s with twenty to thirty animals being seen in every field. Following the arrival of myxomatosis, rabbits were wiped out in the area by 1956 (James, undated).

Since the mid fifties, there has been a slow recovery in the numbers of the species as resistance to myxomatosis has built up within the rabbit population. Rabbits are again a common sight in the Carmarthenshire countryside, especially in the north east, though by no means as abundant as formerly. In many of their original haunts on the coast rabbit numbers have been greatly reduced, so that under grazing now represents a major management problem. Myxomatosis is still occasionally reported, though mortality is rarely as total as that experienced in the 1950s. In 1985, many rabbits were present in the Rhandirmwyn area, but myxomatosis wiped them out within six weeks. Apparently there was quite a smell of rotting rabbits in the area around the Gwenffrwd RSPB reserve, which was the cause of some complaint by visitors.

The rabbit still has some way to go to recover lost ground, particularly in Wales, which has one of the lowest rabbit populations in Britain (Harris *et al.*, 1995). Though the rabbit is once again a familiar sight in the Carmarthenshire countryside, it is unlikely that it will ever reach the populations seen earlier in this century.

BROWN HARE
Lepus europaeus
YSGYFARNOG

A widespread but declining species of grassland habitats.

The brown hare is still widespread, but has undergone a marked reduction throughout Britain since 1960 (Tapper, 1992), a decline mirrored in Carmarthenshire. It is an animal of open habitats, probably introduced to Britain by the Romans (Morris, 1993; Harris *et al.*, 1995).

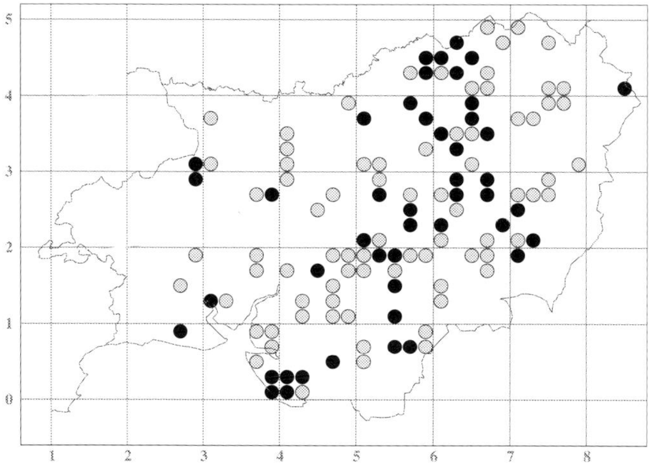

Male brown hares do not defend territories, but guard individual females as they come into oestrus. Aggressive encounters are frequent, but the familiar 'boxing' of hares is actually females attempting to deter the attentions of over eager males. Diseases and poor nutrition also account for much adult mortality, as does a high level of predation by foxes. However, this is unlikely to be a cause of the decline, because high breeding rates of brown hares are enough to counter balance any losses by natural causes. For similar reasons, hunting also has minimal impact on brown hare populations nationally, and may even be indirectly beneficial as it encourages farmers to provide good hare habitat (Morris, 1993).

The first historical record of hares in Carmarthenshire comes from a single tibia found in Ogof Pentywyn, also known as Bone Cave, near Pendine, in 1968 (Davies, 1968). The precise age of this find is not mentioned. A reference to the species in Romano-British times comes in Lloyd-James (1909) which refers to Macsen eating hare caught in the Trefawr area, and supplied to him by a local warlord. The gradual clearance of woodland, and the increase in rough pasture, in the following centuries would have made Carmarthenshire ideal hare habitat.

The decline of the brown hare in Carmarthenshire since 1950 is attested to by many local naturalists and country people, though poorly documented. The species appears to have slipped from view without comment, at a rate that has probably increased since about 1970. Brown hares were thought to be widespread but decreasing throughout Wales shortly after the Second World War (Matheson, 1955). General reports often recorded the species as present but not common, e.g. at the Gwenffrwd RSPB Reserve in the upper Tywi valley (Humphries, 1971). Brunker (*c.* 1950) regarded the hare as 'not common' in the lower districts of Llanegwad parish, but noted numbers breeding in the hills that were forced to lower ground in hard weather. The species may have had a brief pause in its decline when the disappearance of rabbits caused by myxomatosis temporarily removed one competitor. Hares were reported to be more common in 1957 near Rhydargaeau following the extermination of rabbits (Davis, 1957). It was also possible to see brown hares in the Dryslwyn area in the 1960s. Local people reported that up to five animals might be seen in fields alongside the Tywi and near Llandeilo, but they all but disappeared in the early 1970s following an increase in sheep grazing. Brown hares have also become increasingly rare at Gwenffrwd since 1979.

The overall trend has been for the species to disappear from the sheep and dairy country of the west, and is now largely confined to the less intensively farmed areas in the east and north. However, it is still possible to see brown hares with comparative ease at Rhos Pwll y Gawnen near Trelech (where, incidentally, hare coursing was still taking place in the 1980s), whilst much of the surrounding country by sheep to provide suitable habitat. Farther east, they are still relatively common around the foothills of Mynydd Du, the uplands between Pumsaint and Rhandirmwyn, the rides and open glades in Pembrey Forest and the unimproved fields east of the Gwendraeth Estuary.

Changes in agriculture are accepted as the most likely cause of decline, though the precise mechanism is poorly understood (Harris *et al.*, 1995). Brown hares seem to do best in diverse arable landscapes, where a range of food is available to them throughout the year. They fare worse in areas where cattle and sheep are present, probably because countryside used for intensive lamb and dairy production has a reduced diversity of potential food sources (Barnes, Tapper and Williams, 1983). The gradual change in agriculture in the county in recent years, with the change from a mixed farming economy to one based on sheep and dairy may also be responsible for the decline. Studies in Poland have also shown that 44% of leverets were killed in lucerne fields during cutting. Similar mortality may also occur in Carmarthenshire because of silage production, which is cut in a similar manner. Whilst hunting does not have an effect overall, irresponsible gun clubs may tip the balance when brown hare populations are under stress. This reportedly occurred in the Pencader area where, according to local people, such a club virtually wiped out a hare population over four years in the early 1990s.

The brown hare is not a species best served by protected or designated sites. It is an animal of the open countryside, and thus a barometer for the health of our rural environment in general. As a result, the future of the brown hare in Carmarthenshire is largely in the hands of the farming community and those who influence agricultural policy.

The brown hare is a Red Data Book species, but despite this receives no legal protection, apart from the Ground Game Act (1880) and the Hare Protection Act (1911), which restrict certain activities related to hunting.

RED SQUIRREL

Scurius vulgaris

GWIWER GOCH

A much declined species now confined to conifer forests in the north east of the county.

The red squirrel is Britain's only native squirrel, and is still common in Scotland, but is vulnerable in Wales and England (Arnold, 1993). It is primarily an animal of northern conifer forests across most of its European range, but in the past occupied deciduous woodland in the absence of any competing squirrel species (Morris, 1993).

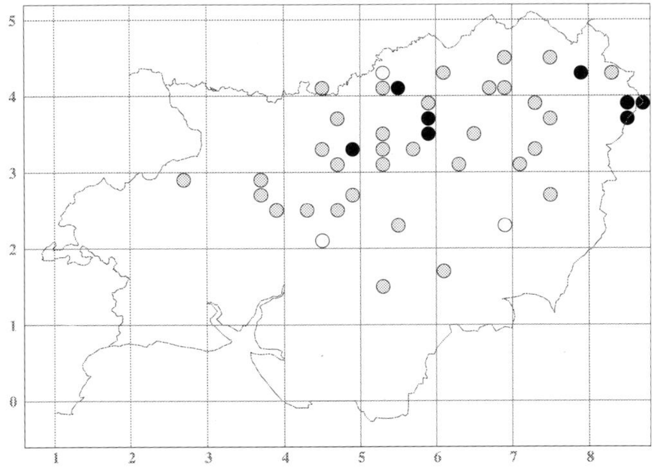

Red squirrels are light weight animals, allowing them to reach the tips of branches, where they spend 70% of their time, to find food. As a result they do not put on fat to carry them through the winter period and consequently are very prone to food shortages. Red squirrels feed opportunistically on a variety of foods including hazelnuts, beech mast, conifer cones, fungi, berries, buds and caterpillars. Owls and pine martens occasionally prey upon red squirrels, but most succumb to starvation and parasitic diseases. Road casualties also claim a few, but are unlikely to have significant effect (Gurnell, 1991a).

The red squirrel has declined severely since the introduction of the grey squirrel from North America, though some decline was evident in the 1920s, before grey squirrels arrived. Grey squirrels occur at higher densities than the native red and are much more terrestrial, spending up to 80% of their time on the ground. As a consequence, habitat fragmentation influences them far less. They are also able to exploit the single most abundant food source in deciduous woodlands: acorns. By contrast the red squirrel does not have a gut which is well adapted to using this food. Finally, grey squirrels put on substantial amounts of fat before winter, and are therefore much less susceptible to starvation in cold weather. All these factors put the red squirrel at an ecological disadvantage, and means that they cannot persist for long in deciduous woodlands once grey squirrels arrive (Harris *et al.*, 1995; Morris, 1993).

In Wales and England, red squirrels only survive in places where large tracts of conifer forest exist, since their low population density and arboreal habits make them vulnerable to extinction in small woodland blocks. They are threatened even here by the planting of deciduous trees along woodland edges and streams. These provide pathways for grey squirrels into the forests, a practice that, ironically, is encouraged by conservationists seeking to 'improve' these habitats for wildlife. Furthermore, since most commercial conifer woodlands are harvested long before the trees produce cone crops for the squirrels, large diverse conifer woodlands appear to offer the only answer (Gurnell and Pepper, 1993).

In Carmarthenshire, with its widespread woodland, red squirrels would have been common, or even abundant throughout the county until about 1950. Barker (1905) regarded the 'squirrel' (the grey squirrel was unknown in Carmarthenshire at that time) as common, and Millais (1904-6) regarded the species as abundant in Wales. Many older residents of the county remember the species as abundant in woodlands throughout Carmarthenshire during the 1930s, in places from which it has long since disappeared, such as near Carmarthen, Cae'rbryn and Llandeilo. One person could remember the chattering of red squirrels in woods near Llandybie in the late 1930s, a site that has long since been obliterated by open cast mining. There is also a specimen skin in the National Museum of Wales from Fforest, obtained in July 1953. Brunker (*c.* 1950) recorded it as quite common in the wooded districts of Llanegwad parish. Panting (1973) recorded that there were 'good populations' in the Pumsaint, Rhandirmwyn and Cilycwm areas in the late 1950s.

Since then, the story of the species in the county has been one of contraction of its range northwards and eastwards. Lloyd (1983), in his review of the status of both squirrel species in Britain, reported the range of the red squirrel in Carmarthenshire contracting from 17 10 km squares in 1959 to 10 in 1971. The last red squirrel in the Llandeilo area was seen in about 1963. The decline in red squirrel numbers may have been exacerbated by the severe winter of 1962-63. Panting (1973) recorded a marked drop in the population at about that time, with animals found dead in dreys. Red squirrels were also common in Crychan Forest in the mid 1960s. Even there the future seemed bleak. Colin Matheson, then Keeper of Zoology at the National Museum of Wales, contacted the Forestry Commission asking if they could obtain a specimen skin from the area before the population completely disappeared (Matheson, 1966). The decline was far from uniform however, with small populations of red squirrels clinging on in suitable habitat whilst they disappeared from surrounding areas. Both Brown (1990), in his survey undertaken at the Dyffryn Conwy Eisteddfod in 1989, and Lloyd (1983) reported sightings of red squirrel from the Lliedi Reservoir area north of Llanelli. However, the veracity of these reports is contested by other local naturalists active in this area at the time. A further record in Brown (1990) from between Cross Hands and Carmarthen for the 1980s is even harder to believe, as the individual would have been well outside the red squirrel range, which was by that time confined to the north and east.

In north Carmarthenshire red squirrels have managed to survive up to the present day. Between 1969 and 1972 it apparently even enjoyed a brief renaissance, with a rise in individuals noted in the north of the county (Anon., 1970; Panting, 1972, 1973). An increase was noted in the Cwmcarfan, Llansawel and Brechfa Forests, where dreys were found along nearby roads and up to four being seen near the Forestry Commission Depot in 1971. There was also a marked increase near Rhandirmwyn, and workers at Llyn Brianne Dam (which was under construction at the time) saw red squirrels daily. Another was seen to run across a road near Ystradffin in September 1971. Lockley (1970) also records that, at that time, it was possible to see the red squirrel at Allt Rhyd y Groes and Gwenffrwd. They also occupied dreys between Cwrtycadno and Rhandirmwyn, chiefly in hazel hedges (Panting, 1973) The species was also present in the coniferous woodland east of Talley until the late 1970s, and a skin obtained in the Cynghordy area in 1972 is now in the National Museum of Wales.

However, from the mid 1980s, all confirmed red squirrel records have come from just three areas. By far the most prolific has been the Sugar Loaf picnic site where animals are seen on a regular basis feeding on rubbish and crossing the nearby road from Llandovery to Builth Wells (Lucas, 1996; Slater, 1987). Records have also emanated from Brechfa Forest, from the Rhydcymerau area, again of animals near a busy road. Connecting these two sites is a third record of an individual brought in dead by a cat near Rhandirmwyn Forest in late September 1995 (Lucas, 1996). The pristine condition of the teeth of this individual, combined with its low body weight suggested that it was a young animal, raising the possibility of a small breeding population in this area. It appears, therefore, that the population in the 1990s is confined to three core areas, based around the large conifer forests of Crychan, Tywi and Brechfa.

The future for this attractive species in the county is dire. The comment in Morris (1993) that the red squirrel may become extinct in Wales by 2010 unless urgent conservation measures are taken is a salutary warning on the wisdom, or otherwise, of casual introductions of non native species. Unless there is a marked reversal in the fortunes of the red squirrel, there will be little that Carmarthenshire naturalists can do except monitor the decline.

GREY SQUIRREL

Sciurius carolinensis

GWIWER LWYD

First recorded in the middle of this century, now abundant in woodlands throughout the county.

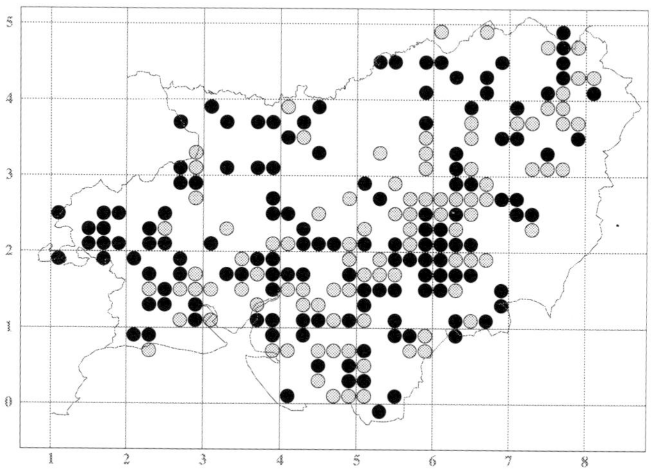

Following its introduction to Britain in the 19th and 20th centuries, the grey squirrel is found throughout most of Wales. Sightings peak in April and June, when squirrels are breeding, and again in October, which may reflect post breeding juvenile dispersal from breeding areas (Arnold, 1993).

Grey squirrels are found in a wide range of habitats associated with trees, such as deciduous woodland, mixed woodland, parks, hedgerows and gardens. They feed primarily on seeds, but have catholic tastes that have contributed to the species' success in Britain. They are in particular more active on the ground than red squirrels.

Grey squirrels raise one or two litters, of two to four young, each year during February and July, in a shallow nest or a hole in a tree. Food supply is the most important factor influencing breeding success. Predation is not important in controlling grey squirrel populations, and human control is similarly ineffective. Large numbers are killed for sport each year in the USA with no effect. Grey squirrels suffer 75% mortality in their first winter, caused by food shortages, but if they survive this period can live up to seven years (Gurnell, 1991b).

The grey squirrel was first introduced into Britain in Denbighshire and Montgomeryshire in the early 19th century (Harris *et al.*, 1995). They were subsequently released in Glamorgan in 1922 (Matheson, 1955), but did not make inroads into Carmarthenshire until several decades later. An anonymous article in the *Llanelli Mercury* ('Deerfoot', 1929) stated that the grey squirrel had not been seen in the vicinity of Llanelli. Brunker (*c.* 1950), who was clearly aware of the presence of grey squirrels in Wales, was quite emphatic that the species did not occur in the Llanegwad area in the early 1950s.

In view of this, the conquest of Carmarthenshire by the grey squirrel appears to have been remarkably rapid. Lloyd (1983), with information from seven surveys between 1931 and 1971, gives the approximate date for the first arrival in the county as between 1945 and 1952, in the easternmost part of Dinefwr. An individual killed at Halfway, some five miles east of Llandovery near the Carmarthenshire - Powys border in 1949 (Evans, 1949) is the first documented county record. Others were trapped at Llandeilo, Dinas Quarry near Brechfa and at Cilycwm in 1950 (Matheson, 1950; Russell, 1950; Slow, 1950; Vaughan, 1950). Local people report other early sightings at Llys Newydd and at Nantgaredig, both in 1953. The relentless march westwards continued in 1956 when the first county sighting to be recorded in *Nature in Wales* was noted when a lone individual was shot at Swiss Valley, Llanelli. Two more were seen at the nearby Upper Lliedi Reservoir the following year (Sansbury, 1956). By 1959, grey squirrels were reported as increasing in the Llandeilo area, and individuals were seen near Taliaris in August and September of that year (Davis, 1959).

Lockley (1962) reports that by 1961 the grey squirrel had completely occupied west Wales, with breeding reported from near Aberystwyth and Pembroke. Many migrant animals were observed in open country such as high ground above the tree line and in rabbit burrows on sand dunes in the autumn period. The severe winter of 1962-63 may have further assisted the species' spread, as there was a marked increase at that time and a corresponding decline in red squirrel (Panting, 1973). The hard winter did not affect the grey, which is hardly surprising given the severity of the winter weather in the species' native area, the north east coast of North America. By 1969, grey squirrels were considered unusually numerous throughout Wales (Anon., 1970) and by 1971 they had recorded the species in every 10 km square in Carmarthenshire (Corbet, 1971). The grey squirrel was here to stay.

Grey squirrels are now abundant in wooded or shrubby habitats throughout Carmarthenshire, from urban Llanelli to the wild upland woods of the north east, and are only absent from the largest and most isolated conifer forests. Road casualties can give an indication of the abundance of this species. For example, in 1996, six freshly dead specimens were noted between Cynghordy and the Sugar Loaf, with a further five newly killed animals three weeks later. When dispersing in autumn, young grey squirrels can even be seen in quite unpromising habitat as they traverse the inhospitable terrain in search of suitable habitat (Condry, 1981). Examples are the low lying coastal belt around Llanelli, an area currently devoid of woodland, and upland locations such as Mynydd y Betws. The grey squirrel is recorded in all months of the year, but there is an increase in autumn when this dispersal takes place, the same phenomenon that so amazed Lockley back in 1961.

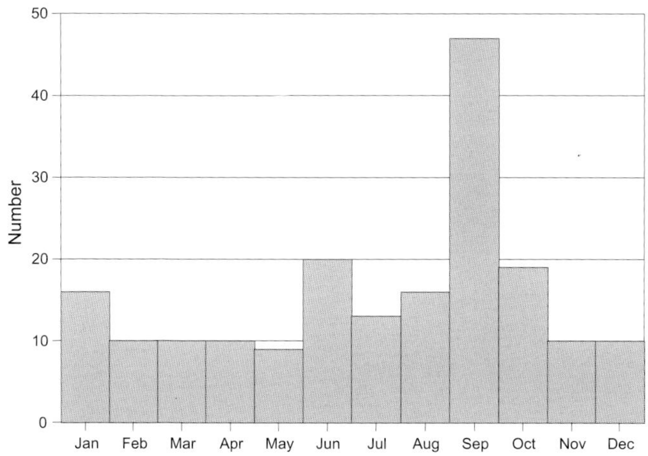

Graph 4 - Monthly Grey Squirrel Records in Carmarthenshire to 1996

BANK VOLE

Clethrionomys glareolus

LLYGODEN BENGRON GOCH

A common species in woodland and hedgerow.

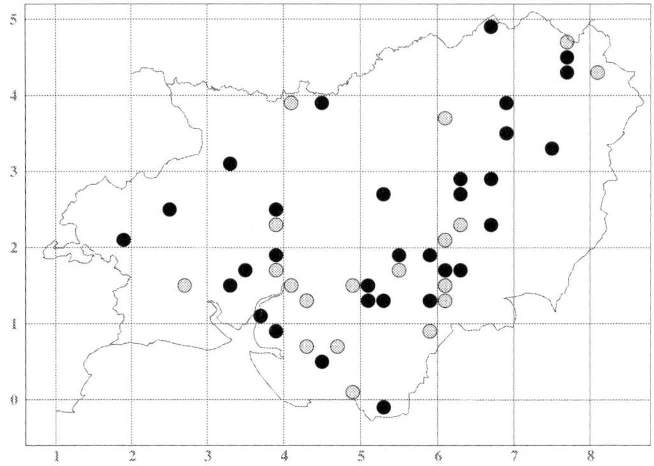

The bank vole is widespread throughout Britain, though it is very much under recorded. Most sightings occur in April and September (Arnold, 1993).

Bank voles are commonest in deciduous woodland with a thick shrub layer, though they are also found in grasslands, conifer stands and hedgerows. It is primarily a herbivorous species, feeding on fruits and seeds and, in winter, dead leaves, though fungi, moss, roots, flowers, grass and even insects and worms have been recorded in the diet. Breeding takes place from March to October with an average of four young being produced every three to four weeks. Young animals born in the spring may themselves breed in late summer, but mortality is extremely high, with 50% dying in their first four months. By early autumn most of the adults have died and the population is mainly made up of young animals. Bank voles that overwinter are born in autumn. Of the fourteen or so recorded predators only owls, stoats, polecats and mink are ecologically important.

Considering the large amount of woodland cover in Carmarthenshire, bank voles are likely to be very common in the county. Their status today is unlikely to be any different from that in the recent past, although the earlier clearance of woodland by mediaeval man must have reduced numbers. The paucity of records reflects the difficulties of identifying small nocturnal rodents: Barker (1906) does not even mention the species, probably because he was unaware of it. Matheson (1955) noted the bank vole as common in suitable habitats throughout Wales, whilst *Nature in Wales* alludes to the species in Carmarthenshire only once, when Sansbury (1956) noted a 'plague' of this and other small rodent species in 1956.

FIELD VOLE

Microtus agrestis

LLYGODEN BENGRON Y GWAIR

A common species of open habitats.

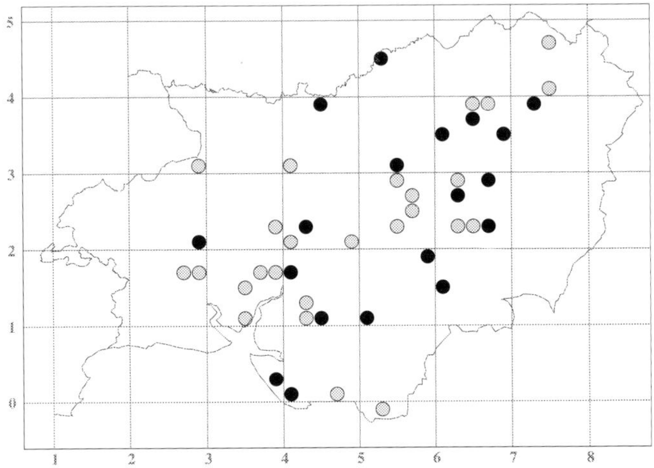

The field vole is abundant throughout Britain (Arnold, 1993). It is found in areas of rough grazing, but also forests, plantations, woods, hedgerows, bogs, dunes, scree, and moorland (Gipps and Alibhai, 1991).

Four to six young are raised in successive litters from March to October, the females becoming sexually mature at six weeks of age. Field vole populations are cyclical, though less so in Britain than on the continent, with periods of abundance and scarcity. Predation helps to dampen these cycles, which are caused by a mixture of intrinsic mechanisms, such as social behaviour and genetics, and external factors such as climate and habitat quality.

The field vole is a pivotal species in the ecology of many communities as they are taken by many predators. Amongst the species with field voles in their diet are grey herons, buzzards, barn owls, tawny owls, long-eared owls, short-eared owls, foxes, stoats, weasels, polecats, pine martens and badgers.

Like other voles, the field vole is very under recorded in Carmarthenshire. Given the amount of suitable habitat, it was probably always widespread and common throughout the county, taking advantage of the mix of woodland and unimproved pasture that until recently covered most of the county. Only passing comment has been made to it by naturalists in the early part of the century; both Barker (1905) and Brunker (*c.* 1950) regarded it as common.

The loss of unimproved grassland in the last two decades will have reduced the population of this species. Field voles were probably more abundant in western Carmarthenshire in the past than they are today if the distribution of predatory species is any guide. Barn owls, which rely heavily on field voles for a major part of their diet, have declined in the county, particularly in the western 'dairy belt'. It is therefore likely that whilst the field vole is still widespread throughout the county, numbers will be highest in the east and south, where the best habitat remains. The field vole does not figure greatly in *Nature in Wales*, but a noticeable increase in numbers in the Rhydargaeau area was recorded in 1956, mirroring a population explosion of the species elsewhere in Wales that year (Davis, 1957; Sansbury, 1956; Cadman, 1957).

WATER VOLE

Arvicola terrestris

LLYGODEN BENGRON Y DŴR

A declining species, mainly confined to eutrophic waters in the south of the county.

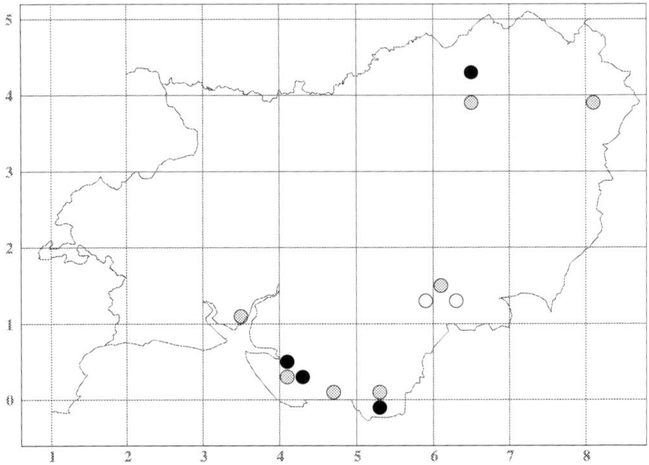

The water vole is still widespread in Britain but is considered to be declining (Arnold, 1993). Its distribution nationally is very patchy, based on core sites of good habitat (Morris, 1993).

Water voles mostly occur on slow flowing rivers and streams that are rich in emergent vegetation. Sites with deep water and steep sided banks are preferred, where they can dig the complex of burrows in which they live. Water voles feed on aquatic and bankside vegetation, but have also been recorded eating insects and molluscs. Breeding takes place between April and September, when they can raise up to five litters of six young. Water voles can live for two winters, but mortality is high, especially amongst young animals dispersing from breeding. Animals of all ages fall prey to grey herons, pike, barn owls, stoats, weasels, foxes, mink, and otters, whilst brown rats may take unweaned young (Boyce, 1991).

There is much anecdotal evidence of a decline in water vole numbers nationally, that is often linked to the appearance of mink. There is a significant inverse relationship between mink activity and water vole activity (Woodroffe, 1994), which suggests that mink do depress vole numbers. Habitat degradation and fragmentation and pollution, together with mink predation, may lead to a highly fragmented distribution. Remaining colonies of voles are then dependent on immigration from outside and can be very vulnerable to local extinctions. Strachan and Jefferies (1993) suggest that 68% of sites with historical records of water vole had lost their occupants by 1989-90 and, on present trends, water voles will eventually disappear from 94% of their former sites.

References to water voles in Carmarthenshire are remarkably few, given the high overall river water quality in the county. Millais (1904-6) regarded the water vole as scarce in south Wales, whilst Barker (1905) noted the species as present near his home on the banks of the Tywi. Brunker (*c.* 1950) regarded it as 'seldom seen but fairly distributed'. They were also thought to be widespread in suitable habitats by Matheson (1955). Water voles were apparently present in the Felingwm area in the mid 1950s, and where Llyn Brianne now stands. One former resident of the Cae'rbryn area could remember water voles inhabiting a brook near Penygroes in the 1930s, and they were also present on the River Loughor near Penybanc in the early 1950s. The extensive marshland that once occupied the lower Gwendraeth Fawr valley must have provided abundant suitable habitat. There can be little doubt that the species has declined markedly in the last 150 years, even without the impact of mink predation.

Most modern records originate from the wetlands in the south of the county. They are regularly recorded at the WWT Centre at Penclacwydd, where abundant suitable habitat, along with fences that protect the wildfowl collection, provides a secure, predator free environment and abundant food. Animals have also been recorded, usually by their distinctive droppings, in ditches and pools outside the centre, in the area dubbed the 'Llanelli Levels', because of its similarity to the Gwent Levels in south east Wales. Water voles also occurred in the Dafen Brook near Llanelli in the mid 1980s, though there is no information on whether they are still present. The small rivers and drains around the Ffrwd Fen area west of Burry Port, and the ditches and unimproved pastures of the lower Gwendraeth valley, provide suitable habitat, although records are few. There are also unconfirmed reports of water voles from the reed beds and marshes surrounding Witchett Pool near Laugharne. Reliable observations are few in this area, a situation exacerbated by the high security that surrounds the military establishment there.

Outside its stronghold in-lying country of the south, water voles have been recorded from the Llansteffan area and near Llandybie. The species has also been recorded at two sites in the north of the county, due to the dedicated efforts of observers who are well acquainted with the field signs of this otherwise elusive mammal. There is also a record of water voles from the Cynghordy area, in 1984, on the Afon Bran. They have not been seen since at this location, although the habitat still appears to be suitable.

Given the paucity of records for the species, the impact of predation by mink cannot be quantified. Considering the widespread nature of mink in the county, Carmarthenshire water voles must come to terms with life with this new predator if they are still to survive in our rivers and wetlands. It should be born in mind that habitat quality is the pivotal factor affecting water vole distribution. To blame the mink for the decline in this species is to abdicate our responsibility to provide the species with a suitable environment in the first place, for water voles can withstand mink predation in good quality habitat.

Habitat creation in areas where the species is still relatively common, such as the planned swannery and wetland development adjacent to Penclacwydd should help to shore up populations in the south east of Carmarthenshire. Elsewhere, the sensitivity of riverine management, together with the interaction between water vole and mink, will probably determine the fate of this most attractive species.

WOOD MOUSE

Apodemus sylvaticus

LLYGODEN Y COED

Abundant in woodland and grasslands throughout the county.

The wood mouse is a widespread and abundant species throughout mainland Britain (Harris *et al.*, 1995).

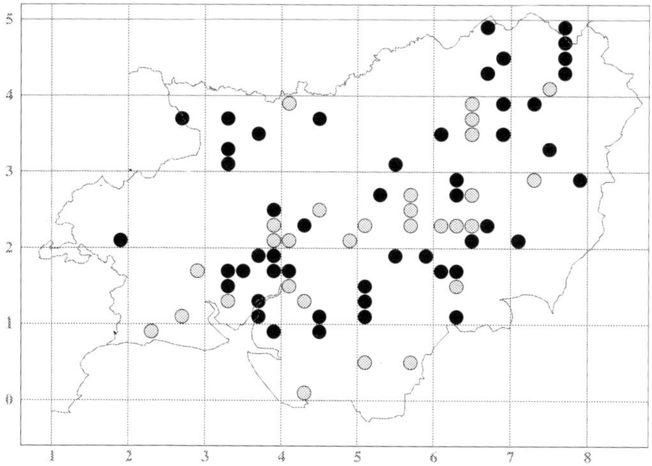

Wood mice are highly adaptable and are ubiquitous in most habitats, being found in woodland, arable land, grassland, heather moorland, blanket bog, sand dunes and hedgerows. Seeds, green plants, and fruit are the main foods, though they may also take buds and invertebrates (Watts, 1968). Between four and seven young are raised in successive litters between March and October. Juvenile survival is linked to the availability of mast crops, and the overwintering population is made up primarily of young animals born late in the summer. Few adults survive two summers. Tawny owls and weasels are the main predators of wood mice, but other owl species, foxes, stoats and badgers also take them (Flowerdew, 1991).

It is likely that the wood mouse has always been common in the county since trees first started to recolonise after the last glaciation. Any gaps in distribution will be due to under recording rather than genuine absence. Barker (1905) noted it as common, whilst Matheson (1955) regarded it as widespread in suitable habitat. Given the abundance of suitable habitat, the future of the wood mouse appears secure.

YELLOW-NECKED MOUSE

Apodemus flavicollis

LLYGODEN FRONFELEN

A localised species, recorded from ancient woodland sites at several locations.

The yellow-necked mouse is one of Carmarthenshire's most enigmatic species. Common where it occurs nationally, its British range is not fully explained. Its distribution is linked to ancient semi-natural woodland, with most records emanating from the Welsh border and south east England. Throughout Britain it has been recorded in just 277 10 km squares, compared with 2,635 for the wood mouse (Arnold, 1993).

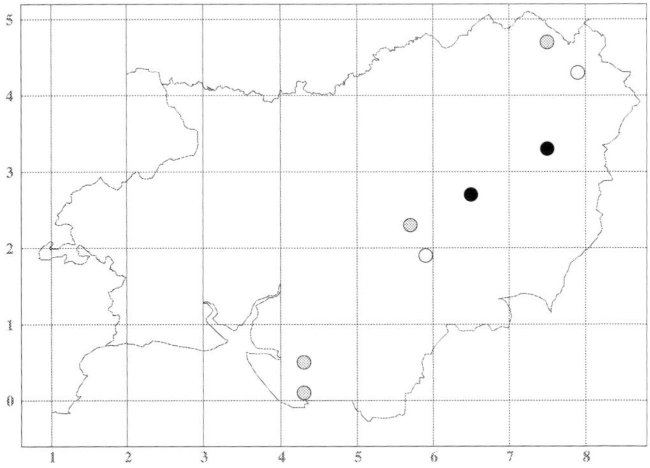

Little is known of the ecology of the yellow-necked mouse, so saying whether it is endangered is impossible (Morris, 1993). The diet is similar to wood mice, though yellow-necked mice may favour woods next to arable fields (Corke, 1977) and may be more arboreal (Montgomery, 1991). Unlike wood mice, they are not found in urban areas, but often come into houses in the countryside. Nearly 50% of records nationally come from inside buildings. Between two and eleven young are raised in successive litters from February to October, though mortality is high and few survive more than one year.

1. Stream near Kidwelly. The overgrown vegetation provides ideal habitat for water voles. *Photograph A. Lucas*

2. Ffrwd Fen from Mynydd Penbre. Harvest mice were confirmed at this site in 1997. *Photograph I.K. Morgan*

3. Talley Pools. Otters have been recorded using artificial holts here. *Photograph I.K. Morgan*

4. Laugharne Burrows. This large sand dune system still has a reasonable rabbit population. *Photograph I.K.Morgan*

5. Carmel Woods. These botanically rich woods are the haunt of dormice. (Photograph I.K.Morgan).

6. Stradey Woods. On the outskirts of Llanelli, this wood host a range of small mammals as well as badgers. (Photograph I.K.Morgan)

7. Near Abergorlech, Brechfa Forest. Home to pine martens and one of the last refuges for red squirrels. *Photograph I.K.Morgan*

8. Overgrown hedges near Llannon. Such hedges are important habitat for dormice. *Photograph A.Lucas*

9. Typical coal field landscape. The Aman valley from Nant Garenig. *Photograph I.K.Morgan*

10. The effect of open cast mining: 'restored' countryside near Pontyberem. *Photograph I.K.Morgan*

11. The Tywi valley from Paxton's Tower. Otters are now regular on the river, whilst fallow deer frequent nearby woods. *Photograph A.Lucas*

12. Unimproved meadow with whorled caraway, lower Gwendraeth valley. Such pastures have abundant small mammal populations. *Photograph A.Lucas*

13. The wild expanse of Mynydd Du. *Photograph A.Lucas*

14. The Pysgotwr valley, north Carmarthenshire. Pine martens may frequent the crags, whilst the wooded valley sides host a range of mammal species. *Photograph I.K.Morgan*

15. Sea cliffs near Wharley Point. Bats are regular in the caves, whilst grey seals are occasionally seen offshore. *Photograph I.K.Morgan*

16. Caves are important wintering sites for bats. The author in one of the Mynydd Du caves. *Photograph A.Lucas*

The first record of yellow-necked mouse for Wales was in 1913 when a specimen was recorded at Rhydyfelin near Aberystwyth (Wright, 1916) some 19 years after it was recognised as a separate species. It was to be nearly 60 years before the yellow-necked mouse was found in Carmarthenshire. Two individuals were caught inside a house at Rhandirmwyn where fruit was being stored in November 1954 (Condry, 1955). Captain H.R.H. Vaughan subsequently sent twenty individuals, all taken between September and March from 1954 to 1956, to Colin Matheson, then the keeper of Zoology at the National Museum of Wales. These remained the only county records until 1961 when E.O. James, the principal at Gelli Aur Farm Institute trapped one at Gelli Aur, followed by two more in January the following year (Matheson, 1965). In 1975, an individual was recorded at Moat Farm near Pembrey, with another being noted near Dryslwyn at about the same time. Subsequently, during the 1980s the species was also found at two other sites; at the Gwenffrwd/Dinas RSPB reserve and at Ty Gwyn near Pembrey, both close to the existing records at Moat Farm and Rhandirmwyn.

These remained the only records for yellow-necked mouse up to the end of 1996. An attempt by the author to rediscover the species at Gelli Aur using Longworth traps in the summer of 1996 proved unsuccessful. However, the county's cats have proved more effective surveyors for yellow-necked mice when, in 1997, records were confirmed at two new locations that linked the two main sites in the north east and south of the county. Individuals were brought in by cats at Ystrad House near Llandovery and near Manordeilo, producing a line of records from Rhandirmwyn in the north east to Pembrey in the south.

Matheson speculated that, given the number of individual records emanating from Carmarthenshire, the yellow-necked mouse probably had a widespread, if scattered, distribution in the county. The shortage of recorders of small mammals, together with the difficulties of separating the yellow-necked mouse from wood mouse has no doubt contributed to the paucity of records. Corke (1977) analysed the distribution of the two species and produced a map of the probable range. This suggested that the species was rare or scattered throughout Carmarthenshire, a conclusion that appears to be supported by existing records.

The yellow-necked mouse's link with ancient woodland, and its distribution centred on southern England and the Welsh Marches, can be compared with the common dormouse which has a similar range. The situation regarding the yellow-necked mouse in Carmarthenshire is very similar to that for the common dormouse prior to around 1990, with a few scattered records. Subsequently, intensive survey efforts have shown the common dormouse to be widespread in the south of the county. Given the amount of available ancient woodlands in the valleys around Carmarthenshire, the yellow-necked mouse may prove to be far more common than is currently supposed. A determined individual, armed with appropriate traps, could do much work to establish the true status of this mysterious species in the county.

HARVEST MOUSE

Micronomys minutus

LLYGODEN YR ŶD

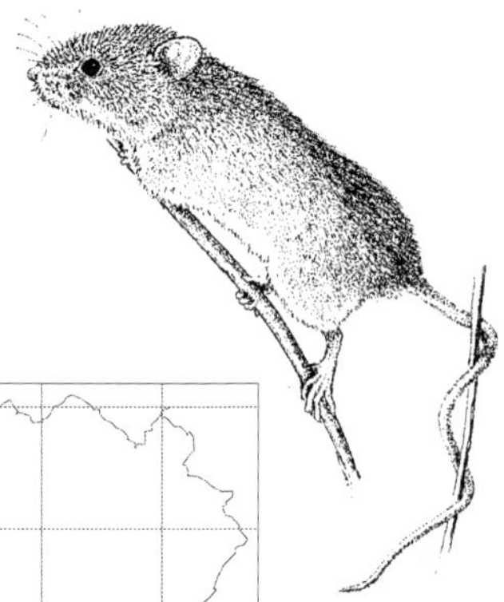

Status uncertain. Only confirmed at one location.

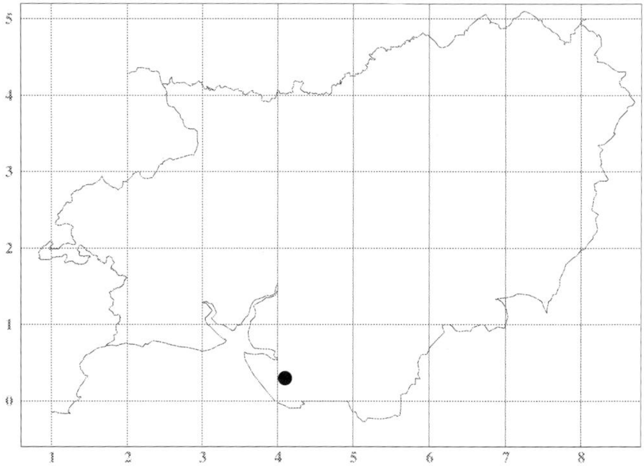

This tiny rodent is the newest addition to the Carmarthenshire mammal list. Throughout Britain, it is found in areas of tall vegetation such as long grass, reed beds, grassy hedgerows and, of course, cereals. It is widely distributed in England, with Welsh records from Gwent, Glamorgan and Pembrokeshire (Arnold, 1993).

Harvest mice feed on grass seed and insects, which they glean by climbing grass and other vegetation. It is thought that the harvest mouse's small size is an adaption to this lifestyle. Harvest mice have a spectacular breeding rate, with females breeding at only 35 days old and producing a litter of five young after only 17 to 19 days gestation. Mortality is correspondingly high, with harvest mice being vulnerable to low temperatures as well as being predated by 'any predator that walks runs or flies' (Perrow and Jowitt, 1995).

The presence of harvest mice in Carmarthenshire has long been suspected. Indeed, the species was included in Appendix 1 of a preliminary draft of *Mammals in Carmarthenshire.* An attempt was made to discover the species in 1996 when traps attached to bamboo canes as well as tennis ball nest sites were tried at several locations in the county. Harvest mice were confirmed in Carmarthenshire in 1997 when two nests were discovered at Ffrwd Fen. These remain the only records, although there is an unconfirmed report of harvest mice near Llandovery in 1968.

The Carmarthenshire record provides a link in the range of the harvest mouse in south Wales, joining the Pembrokeshire and east Wales records. It is quite possible that this diminutive species may be more widely distributed in Carmarthenshire. No doubt other reed beds, such as Witchett Pool, as well as rhos pastures throughout the county would repay further study.

HOUSE MOUSE

Mus domesticus

LLYGODEN FACH

A common but under recorded species in urban areas.

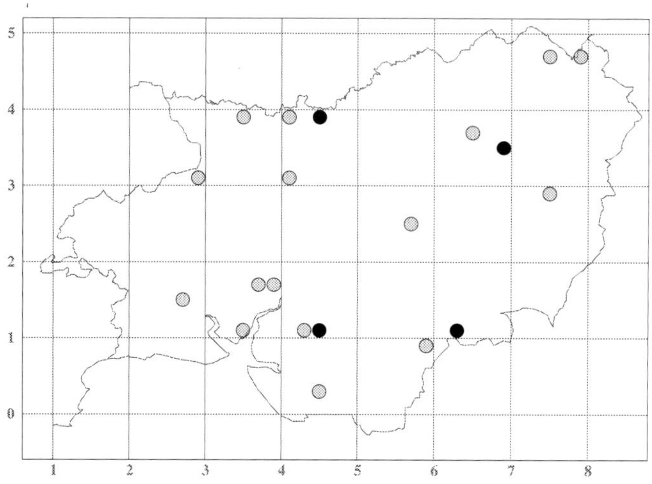

The house mouse is ubiquitous throughout Britain, but very under recorded (Arnold, 1993).

The house mouse is thought to have originated in rock crevices in Iran and the south west of the former USSR, a habitat which pre adapted it admirably to life with humans. It is now found in a range of habitats but is rare in woodland, where it is excluded by other small mammals. House mice mainly feed on cereals, though insects have also been recorded in the diet.

Five to eight young are raised at monthly intervals throughout the year when living indoors, or in the summer when outdoors. The mean life expectancy is 100 days, but mice living with humans may live up to two years. Owls and rats predate house mice, but the main factor influencing survival is cold weather (Berry, 1991).

Little information is available on the house mouse as it comes to the attention of environmental health officers much less frequently than the brown rat, most people dealing with infestations themselves. However, it is likely to be common throughout the county, apart from the most rural areas, and is probably present in every 10 km square.

BROWN RAT

Rattus norvegicus

LLYGODEN FFYRNIG

A common but under recorded species in both rural and urban locations.

The brown rat is widespread and ubiquitous, but like that other commensal rodent, the house mouse, is very under recorded. The species probably originated in the eastern former USSR or China, but spread in the 18th century, replacing the ship rat. It probably arrived in Britain around 1728 in shipping from Russia, and not, as the scientific name suggests, from Norway.

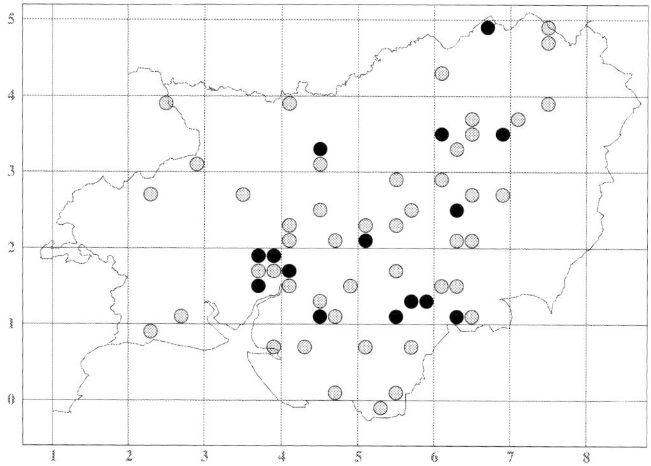

Brown rats are highly adaptable, found in farms, refuse tips, sewers, hedgerows and arable land. It has a varied diet, including meat, fish, bones, cereals, root crops, weeds, and has even been recorded to eat *Spartina* and crustaceans in coastal areas (Taylor, Fenn and McDonald, 1991). Breeding is continuous in suitable environments, otherwise taking place in summer and autumn.

Young rats are eaten by a range of predators, but older animals are too aggressive to have many enemies. Few, however, live more than one year, as food availability governs populations. In areas where the population is saturated, juvenile mortality may be 99%.

After their arrival in Britain, brown rats will have become common in Carmarthenshire, although improved standards of hygiene and vermin control mean that it will be less common than formerly. Davies (1858) reported that rats, presumably brown rats, were common on the banks of the Tywi and that *in winter* (they leave) *the river side to enjoy a comfortable retreat in the barns and out houses of the farms adjacent which they harass, causing not a little mischief among the corn stacks.* Matheson (1931) noted that, in the late 1920s, brown rats were breeding at a rubbish dump a quarter of a mile from Carmarthen but, following the erection of a refuse destructor, few were seen.

Hill (1961) records the brown rat as being widely dispersed in urban and rural Wales, having ousted the ship rat in the previous 150 years, though eradication programmes had, by that time, made it difficult to find large o survive despite all the efforts at control and is probably present in every 10 km square in the county. The responsibility for brown rat control falls to the environmental health department of the local authority. By far the most complaints regarding this species originate from the Llanelli area. This is largely because of the higher human population, therefore more potential complainants, and also because of the greater concentration of sewers in this urban area. Up to two thousand enquiries are recorded in the Llanelli area annually, mostly in the winter when rats come into buildings for warmth. Rat problems on farms are also more likely to be dealt with by the landowner concerned without ever coming to the attention of the authorities. Long term trends in brown rat population are difficult to discern, because of the complicating issues of weather and sewer building. People are also more likely to complain now than in the past! Considering the abundant habitat and food, both in agricultural and urban locations, it is likely that this most unpopular member of the Carmarthenshire mammal fauna will be with us for many years to come.

COMMON DORMOUSE

Muscardinus avellanarius

PATHEW

A local species, confined to the woodlands and hedgerows of the south and east.

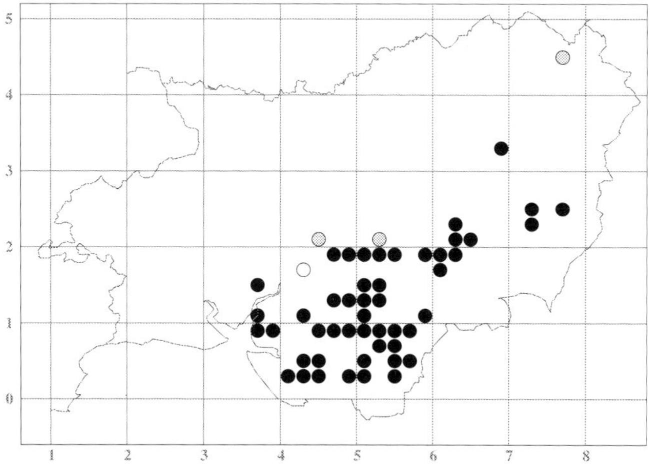

The common dormouse is widespread but local in suitable habitat south of a line from mid Wales to Suffolk, with isolated records from areas further north. It is a declining species that is very prone to local extinctions caused by inappropriate woodland management. The discovery of a thriving population of dormice has been one of the most significant events in the study of mammals in Carmarthenshire in the last decade.

Dormice are found in deciduous woodland with abundant secondary growth, coppice and also hedgerows, especially those that are species rich. The physical structure of the habitat is important. Since they spend nearly all their time in trees, aerial pathways of interconnecting branches are essential to allow animals to move around their territories. They are almost entirely active at night, and are specialist feeders on a succession of foods such as flowers, pollen, insects and nuts. Hazel nuts in particular are a critical food in enabling the species to put on weight prior to hibernation.

Dormice live in pairs or one male with two females, and range for about 0.5 ha. Population densities are low, generally three to ten per hectare, one tenth of other woodland rodents such as wood mice and bank voles (Bright and Morris, 1993). During the summer they occupy nests above ground, of woven vegetation, in which they raise their young. They produce a litter of about four young in mid June, but in years of poor weather or low food availability this may be delayed until August or even September. By contrast, hibernation takes place in a tightly woven nest on the ground, where temperatures are more stable. Dormice can also enter a dormant state during the summer when food is short, known as summer torpor. Though they are slow breeders, they can live for up to four years in the wild, compared with about a year for other rodents of the same size.

The common dormouse can therefore be regarded as animal with very specific habitat requirements. Because of this, it is very susceptible to local extinctions. Its inability to disperse over long distances means that dormice often occupy 'islands' of suitable habitat separated by tracts of inhospitable terrain. Once lost from a site, are unlikely to return unaided (Bright and Morris, 1990).

The first reference to the common dormouse in Carmarthenshire is in Phillips (1885) who regarded it as rare in parts of Carmarthenshire. However, the species appears to have been well known to country people, as the common dormouse appears in sayings such as *cyn dewed a phathew* (as fat as a dormouse) and *cysgu fel pathew* (to sleep like a dormouse) (Condry 1981; Mathew 1997). The existence of a local dialect name - pathor - which was apparently used in east Carmarthenshire and Glamorgan, suggests that dormice were familiar animals in those areas (Griffiths and Jones, 1995; Mathew, 1997). Millais (1904-6) thought the common dormouse probably occurred in most parts of Wales and quoted Phillips' record, whilst Barker (1905), described it as 'rather local' and records finding a 'colony' near his home, 'Oaklands', at Cwmffrwd on the outskirts of Carmarthen. Brunker (*c.* 1950) notes only one record from the Llanegwad area, and only four pre 1990 records, from St. Ishmaels near Ferryside, Lliedi Reservoir, Capel Dewi and Tir Philip near Llanarthne are recorded in Mathew (1992). The species had also been recorded in the north of the county by Hurrell and McIntosh (1984) during a Mammal Society survey in the mid 1970s. The fact that this elusive species had been present in Carmarthenshire had been known for more than a hundred years, even if its exact status was unknown.

The explosion of interest in dormice dates from the discovery of an individual killed by a cat near Llannon in 1991 (Mathew, 1992). A search for the characteristically chewed hazelnuts which suggest the presence of dormice revealed signs at several sites in the Gwendraeth Valley area, with a live sighting near Four Roads in September 1992 (Lucas, 1993a). A Countryside Council for Wales survey in February 1993 found a further fifteen sites south of a line from Tumble to Kidwelly (Mathew, 1993), indicating that, in this part of Wales at least, the common dormouse is well established (Lucas, 1993b).

Subsequent survey work by the small number of 'dormouse enthusiasts', aided by the publicity generated by 'National Dormouse Week' in 1993 raised the number of records by December 1996 to 66. The vast majority of these records have been in the form of chewed hazelnuts. However, a hibernating individual was found in a shed near Mynydd y Garreg in 1994 and another observer has reported seeing mice-like animals at night in bushes in the woodland close to his home near Crwbin. Another hibernating individual was discovered in a leaf pile near Llansadwrn in 1996. Llannon, in particular, has become something of a 'hot spot', with many records in this area as a result of the enthusiasm of local naturalists. Nest boxes designed for use by dormice, with the access hole on the back of the box, have been erected at several locations in the county. They were first used 1995 when young were raised in boxes in hawthorn and hazel scrub at Llwyn-teg near Llannon. A Wildlife Trust reserve has now been established in the area, with hedgerow management designed to encourage the species. A cat has also brought in two dormice near Llwyn-teg in 1996. The animal has now been deported to English exile! The same year a common dormouse also fell victim to a cat near Llandeilo.

Searches elsewhere in the county have failed to produce additional records, even in what appears to be suitable habitat. These include south west Carmarthenshire near Llanddowror, the Duad valley near Cynwyl Elfed and the Cych valley near Cwm Morgan, while Clarke (1997) did not rediscover the common dormice found near Rhandirmwyn in 1975. The bulk of the Carmarthenshire population therefore appears to be confined to a triangle between Llanelli, Llandeilo and Carmarthen.

Habitat choice by dormice in Carmarthenshire has been thoroughly reviewed by Mathew (1997) who undertook a survey of all known sites in 1996 and measured the habitats using a number of criteria. The 66 sites surveyed are spread over eleven 10 km squares, with SN50 in the south east having the most (29% of all records). The species is primarily one of lowland woods; 70% of sites are in woodlands of some kind with 8% each in hedges and scrub, and 16% in other habitats such as gardens and riverine trees. Some 41% of sites are in ancient woodland. However, it should be noted first that much initial searching was directed towards this habitat, as it was thought the most likely to produce success, and secondly that woods below 2 ha do not qualify on the ancient woodland criteria. Despite these caveats, dormice in Carmarthenshire appear to be using different woodland types to other areas studied in Britain such as Herefordshire, where most records come from relatively large ancient woodland sites (Bright *et al.,* 1994). Most sites are also close to watercourses. Ancient woodlands, particularly associated with the river valleys of the Morlais, Lliedi, Gwendraeth Fawr and Sawdde are important both as a habitat and as corridors to allow dispersal of immature animals.

Woods occupied tended to be small, with a median size of 7 ha (the smallest near Llannon being only 0.5 ha) and were often connected to nearby woods and scrub by hedgerows, riverine trees or, in several cases, by scrub alongside disused railways. Oak tended to be the dominant canopy species but, significantly, woods usually had a thriving understorey, mostly of hazel and were ungrazed in 72% of cases. Grazing stock is thought to be inimical to dormice because it damages the understorey and tramples hibernating animals. A significant number of sites (27%) were in coniferous woodland replanted on ancient woodlands. Although conifers are usually thought to be detrimental to dormice, they appear to be able to survive in them, provided a significant deciduous component, especially of hazel, remains. The typical Carmarthenshire common dormouse site is therefore a small woodland in the south east, usually ungrazed, with abundant hazel, and linked by scrub or hedgerows to other similar woods nearby.

It has been estimated that an isolated woodland would need to be at least 20 ha in extent to support a self-sustaining population of dormice (Bright *et al.*, 1994). Only about a quarter of sites with dormice are of this size in Carmarthenshire. The network of hedgerows and other linear features are critical in allowing dormice to recolonise sites when population levels in any one area are low. It is also worth noting that at least 50% of the county records emanate from what could be loosely described as non ancient habitats, including scrub, secondary woodland and deciduous woodland on the edge of conifer blocks. For a species noted for its inability to disperse over areas of unsuitable habitat, the continuing survival of a complex network of hedgerows is important in allowing dormice to colonise new habitat.

The lack of records, even from habitats that appear suitable has been noted by Bright (1995). Why are dormice seemingly absent from much of north and west Carmarthenshire? There are several possible reasons, which probably all contribute to the paucity of records from this area. First, the Afon Tywi would effectively isolate populations to the north from their 'heartland' in the south and east, making them more vulnerable to extinction. Secondly, the habitat may not be suitable, as this area has been subject to much greater agricultural 'improvement', hedgerow removal and clear felling of woodland (Linnard, 1982). This may have caused extinctions in isolated woods which dormice have been unable to recolonise. Thirdly, the more extreme climate of upland Wales, particularly the wet summers, may affect common dormouse breeding success. Finally, there is always the possibility that dormice are indeed present, but that recording efforts have failed to locate them, and more work is needed.

In late 1995, a survey attempting to find harvest mice at Ffrwd Fen, using plastic traps mounted on poles, trapped two dormice, an adult and a juvenile, in the reed bed some 100m from hawthorn scrub (Lucas, 1996). This habitat is not noted as one usually being utilised by the species, although the occurrence is not unprecedented, for dormice have been found in similar habitats in south east England (P. Morris, pers. comm.). Whether the species uses reed beds on a regular basis is unknown. However, they may actually be quite a good habitat for dormice, affording both shelter and food in the form of seeds and aphids.

The common dormouse has taken its place alongside other species associated with the large hedgerows and unimproved fields of the Carmarthenshire coalfield, such as the marsh fritillary butterfly. The conservation of the common dormouse, a symbol of the diverse landscape of south east Carmarthenshire, should be a major priority.

UNIDENTIFIED WHALES

Cetacea *spp.*

MORFILOD

A brief note on whales in general is appropriate because, as cetaceans are so difficult to identify, some anecdotal records refer to unidentified individuals.

The earliest documented record of whales off the shore of the Carmarthenshire coast is a report of three having been harpooned in Carmarthen Bay, reported in the *Llanelli Guardian* of 14 August 1879. Barker (1905) also reports a whale ... *of what species I cannot say* ... coming ashore ... *some years back* ... near Ferryside.

One of the most remarkable reports comes from the *Llanelli Mercury* of 29 August 1907. Given the efforts made nowadays to coax stranded cetaceans back to the sea, the reception that awaited the 'Llangennech Whale' is most illuminating, and is worth quoting at some length:

> *'WHALE CAPTURED NEAR LOUGHOR A great deal of excitement: A whale of considerable dimensions was caught in the River Loughor on Friday, between Loughor and Llangennech. It was observed in the water near Pencoed Ucha Farm first of all ... about 6.30 in the morning. ... (and) ... preparations were made for its capture. But this proved a very difficult task and afforded a great deal of excitement. A gun was obtained and four shots fired at the monster, seemingly with not much effect. Another four shots were aimed at it, and eventually, after the eighth shot had been fired, the huge fish was seen to reel slightly on its side. With further assistance the men decided to approach it, when it was seen that the whale was still alive. Pick axes were brought, and with a great deal of precaution they were pierced into its side. This was continued for about three hours during which it bled profusely. At about 9 o'clock, it succumbed and with considerable difficulty was dragged to the bank.*
>
> *It measured 16 feet 8 inches and 8 feet round the gait and its weight is 1 ton 15 cwt. How the whale got so high up the river ... has been the subject of much conjecture. At the time of its discovery it was in a good depth of water ... sufficient for it return back on its journey.*
>
> *In the afternoon it was brought down to Llanelly in a waggon drawn by two horses and was taken to the market to be weighed. On the Friday and the Saturday it was exhibited at the fete and gala at Stradey, where it was seen by hundreds of people.'*

One of the most striking things about this account is the breathless excitement of the prose, completely devoid of any guilt for having killed a whale, which probably could have returned to the open sea, and in a most gruesome fashion. It is to be hoped that future stranded whales do not meet an equally grisly fate!

Hope Jones (1974) provides a summary of cetacean strandings in Wales up to that time. Summaries are also periodically produced by the Natural History Museum, London, but much of the following data has been supplied from a database of strandings compiled by Dr David George of the Marine Biological Services Division of the museum. Access to this information is gratefully acknowledged.

MINKE WHALE

Balaenoptera acutorostrata

MORFIL PIGFAIN

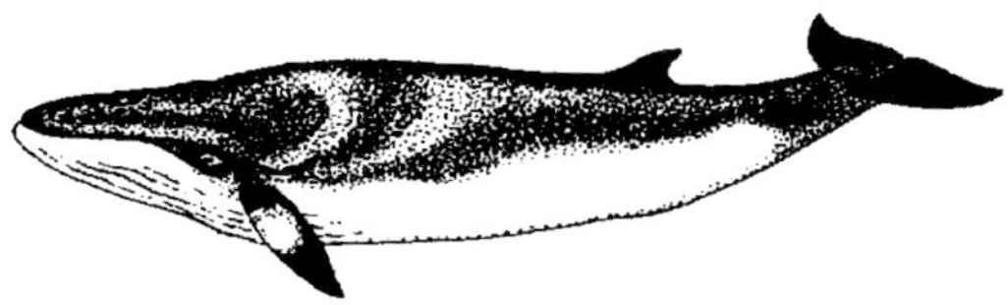

Two stranding records.

The minke whale is the commonest rorqual species found around Wales, the nearest main concentration being found off the south west coast of Ireland. The smallest of the baleen whales, the family that includes the blue whale, it reaches a length of about 8 m and a weight of approximately 10 tonnes. Minke whales feed on fish, euphausiids and pteropods, and have been recorded living up to 50 years (Evans, 1991).

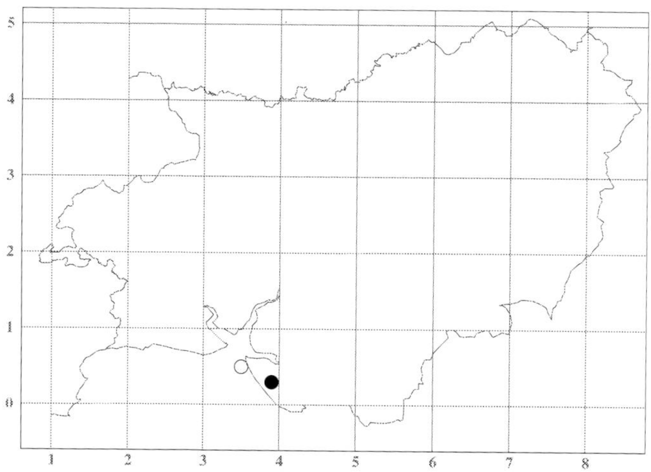

An individual was reported stranded on Cefn Sidan on 22 July 1969 (Hope Jones, 1974). A second, a female possibly ten years old and measuring 8.36 metres, was found at the same locality on 3 January 1991.

NORTHERN BOTTLENOSE WHALE

Hyperoodon ampullatus

MORFIL TRWYN POTEL

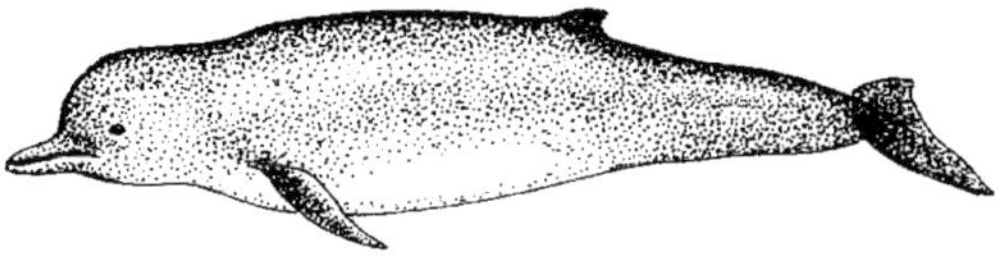

One stranding record.

This whale is found in temperate and Arctic waters in the Atlantic. It is considered rare in Welsh waters but is occasional off the south west coast of Ireland (Evans, 1991), although whaling has depleted the population. The main food item is squid.

There is a single record; one washed up at Burry Port on 4 September 1918 (Hope Jones, 1974).

HARBOUR PORPOISE

Phocena phocena

LLAMHIDYDD

Probably regular in Carmarthen Bay, with many recorded strandings.

The harbour porpoise is our smallest cetacean species, and is widely distributed in the waters of the North Atlantic. It feeds on a variety of fish species, and also some crustaceans and cephalopods. Breeding takes place between May and August, with most births taking place in the shallower coastal waters. The harbour porpoise is regularly recorded in Welsh waters, particularly off the Pembrokeshire coast, with the nearest main population being off the south west coast of Ireland (Evans, 1991).

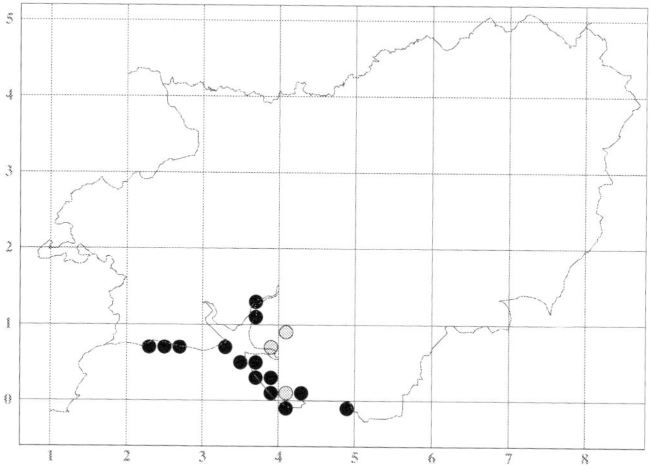

The harbour porpoise is by far the most common cetacean in Carmarthenshire waters, with more strandings than all other species combined. Little is known of its true status. However, following the increased effort at seabird and cetacean recording in Carmarthen Bay as part of research monitoring the effects of the 'Sea Empress' oil spill in 1996, it has become clear that the Bay is an important location. It is also regularly recorded by birdwatchers off Worm's Head, Gower, which overlooks Carmarthen Bay, and in September 1996 was also observed off Marros in the south west of the county. Unsurprisingly, there are few historical references for a species which would have gone largely unnoticed by earlier naturalists, although Barker (1905) regarded this species as common in Carmarthen Bay.

Harbour porpoises show a strong periodicity in the timing of strandings, with 60% of such events occurring during the three summer months. This follows a pattern observed throughout the British Isles which found that, although porpoises are seen at all times of the year, there is a distinct seasonal peak between July and October. This coincides with the period during which most young are born, when harbour porpoises are known to move to coastal waters.

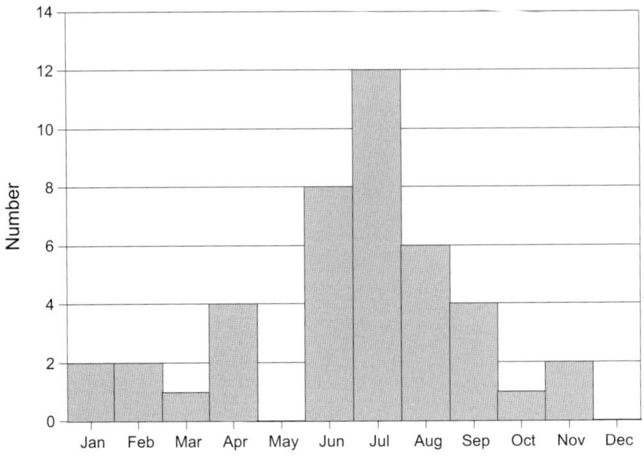

Graph 5 - Monthly Harbour Porpoise Strandings on the Carmarthenshire Coast to 1996

There is also a strong bias in the locations of strandings, with most occurring along the Pembrey/Cefn Sidan area. This almost certainly reflects the prevailing wind from the south west, which would tend to funnel any dead porpoise washing into Carmarthen Bay towards this area of coast.

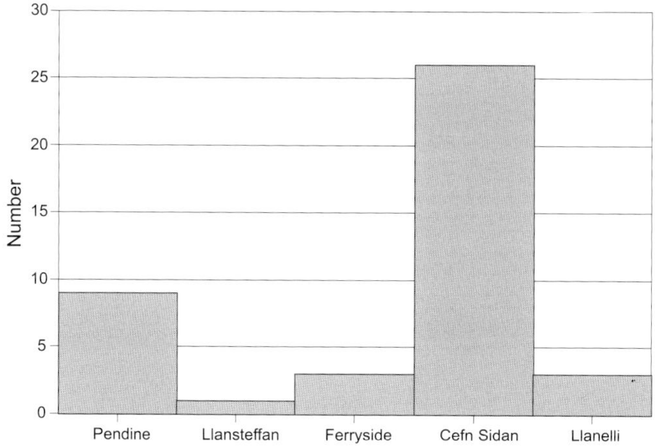

Graph 6 - Harbour Porpoise Stranding Locations on the Carmarthenshire Coast

COMMON DOLPHIN / STRIPED DOLPHIN

Delphinus delphis / Stenella coeruleoalba

DOLFFIN CYFFREDIN/ DOLFFIN RHESOG

There are two records of sstranded indivuals which could not be definively ascribed to one species - one at Cefn Sidan, 9 November 1986, and a second, nearby, on 31January 1988.

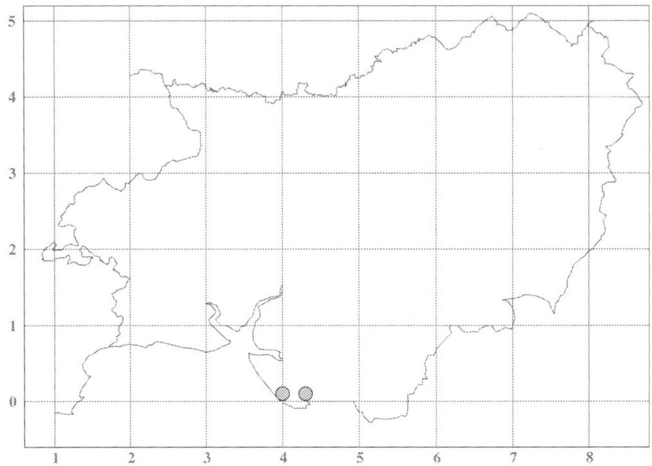

COMMON DOLPHIN

Delphinus delphis

DOLFFIN CYFFREDIN

Occasional off the Carmarthenshire coast.

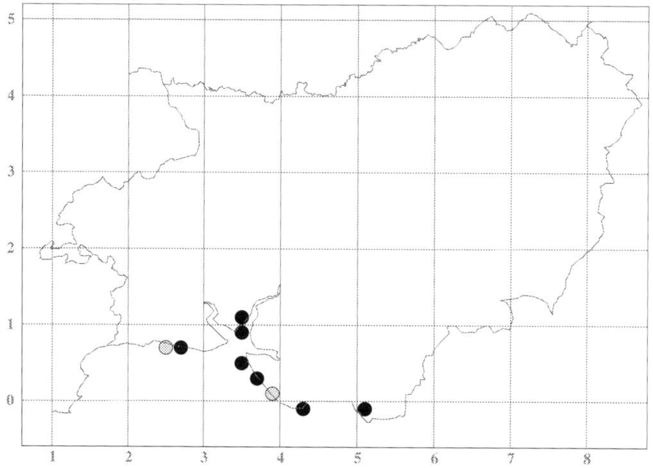

This small, slight dolphin is found worldwide. It is considered scarce off the Carmarthenshire coast, although it is regular in nearby Cardigan Bay. Most sightings occur from July to October, when there is believed to be a slight onshore movement. The diet consists, for the main part, of a variety of fish species and squid (Evans, 1991).

The common dolphin is, after the harbour porpoise, the second most common cetacean recorded on Carmarthenshire beaches. The eleven individuals recorded since 1982 have been widely spread along the coast of the county, and have been found mostly during the colder months of the year.

STRIPED DOLPHIN

Stenella coeruleoalba

DOLFFIN RHESOG

Two stranding records.

The striped dolphin has a worldwide distribution, but is considered rare in British waters, although it is regular off the west coast of France. Most British sightings have occurred off the south west Atlantic and Irish Sea coasts. There has been an increase in records in recent years, although the pattern is clouded by difficulties in differentiating between this and the

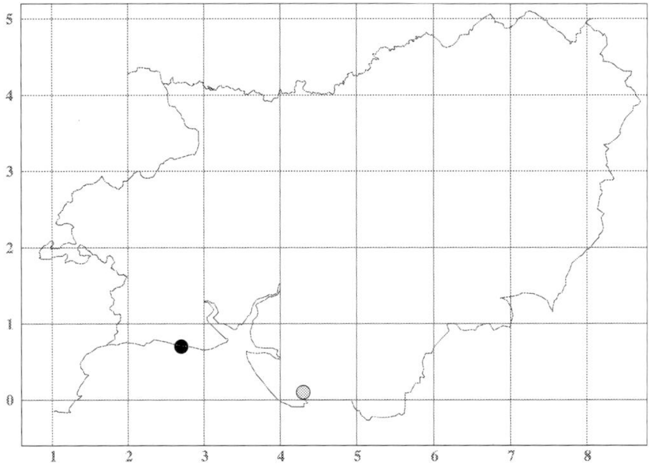

previous species. Like the common dolphin, the species feeds on fish and squid (Evans, 1991).

The two Carmarthenshire records are of an individual found dead at Cefn Sidan on 29 December 1983, followed by a second found dead at Pendine on 22 January 1990.

BOTTLE-NOSE DOLPHIN

Tursiops truncatus

DOLFFIN TRWYN POTEL

Two stranding records.

This essentially inshore species is occasional off the Carmarthenshire coast, the nearest main concentrations being in the Irish Sea and off the coast of western Ireland and Cornwall. The main items in its diet are fish and squid.

Hope Jones records this species in every coastal county in Wales, though he does not give

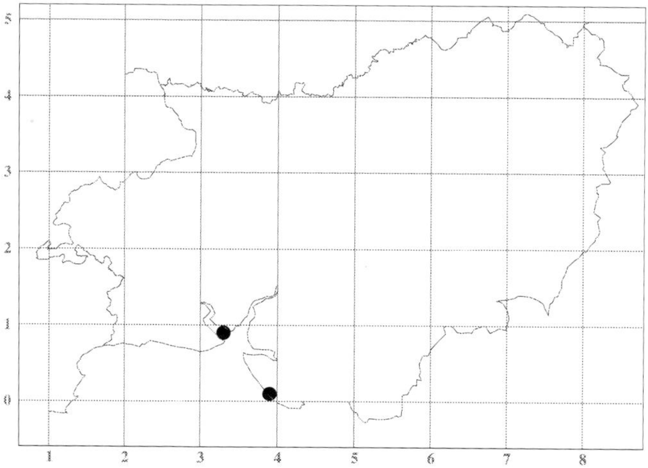

any figures for Carmarthenshire. There are two county records; one found dead at Ginst Point in May 1990, and a second, which suffered a similar fate at Cefn Sidan on 2 August 1992.

FALSE KILLER WHALE

Pseudorca crassidens

COEGLEIDDIAD

Three county records.

False killer whales are found world wide, mainly in tropical waters, feeding on squid, fish and dolphins. In 1927, 150 were stranded at Dornoch in Scotland, 25 in south Wales in 1934, followed by 75 in 1935. There have been no further strandings since that year, although there have been four sightings since 1976; three off Scotland and one off Cornwall (Evans, 1991).

There are three separate records for Carmarthenshire (Hope Jones, 1974; Nicholson, 1996); one, Pwll, 19 May 1934; one at Pembrey on 30 May 1934 and two at Burry Port, on 7 May 1934. At the same time, 21 were stranded across the Burry Inlet at Llanmadoc, Gower on 8 May 1934 (Matheson and Cowley, 1934).

KILLER WHALE

Orcinus orca

LLEIDDIAD

One record.

This powerful and striking species is found worldwide from tropical to polar seas. Evans (1991) regarded it as occasional in Carmarthenshire waters, the nearest concentration being off the south west coast of Ireland. Killer whales have a varied diet including fish, squid, marine mammals, turtles and even birds.

A killer whale was washed up on a beach near Burry Port in September 1971. It was thought that the whale had become stranded whilst chasing mackerel into the gently sloping waters of the Burry Inlet (Anon., 1971).

LONG-FINNED PILOT WHALE

Globicephala melas

MORFIL PENGRWN

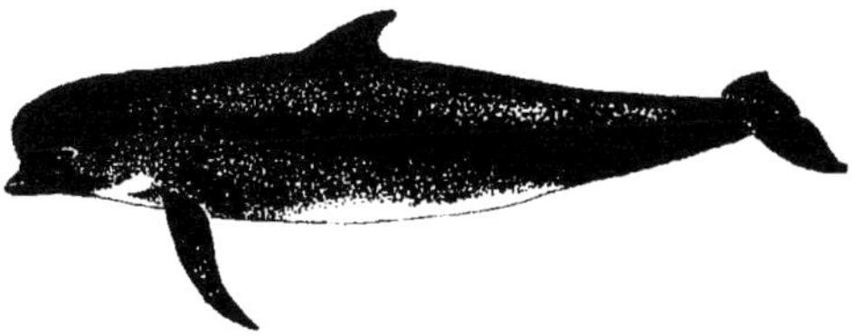

Two county records.

Long-finned pilot whales are widespread in temperate regions, and are considered occasional off the Carmarthenshire coast. The nearest main concentration is off the south west coast of Ireland. The main food item is squid, but they also take fish (Evans, 1991).

There are two records; one stranded at Cefn Sidan on 24 January 1973 (Hope Jones, 1974) and, remarkably, a second near the same locality on 23 January 1974.

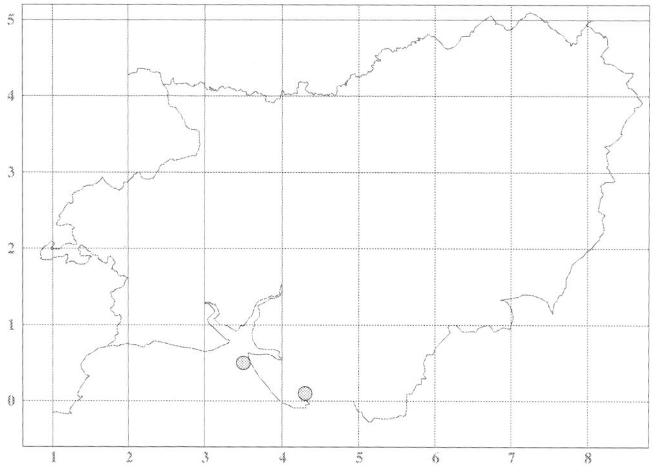

FOX

Vulpes vulpes

CADNO

Widespread and common in most habitats, including urban areas.

Foxes are widespread on mainland Britain and Ireland, and have been recorded up to 1300 m in Scotland.

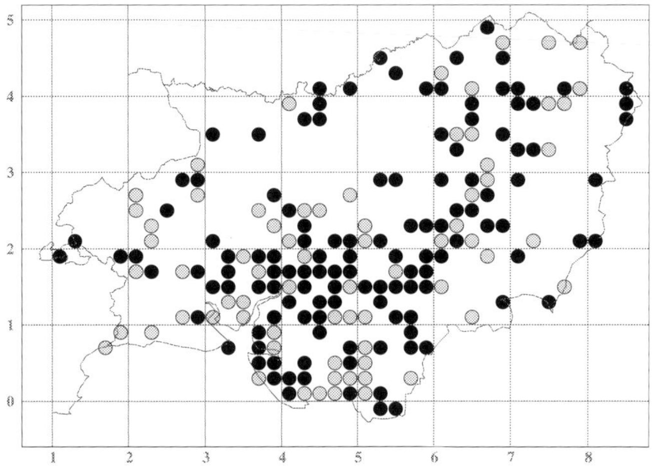

The fox is successful because it is highly adaptable and unspecialised. They are found in woods, farmland, hills, dunes, conifer plantations and urban areas, where they live in family groups whose territory size depends on habitat quality. In mid Wales, territory sizes of one pair per 1.2 - 4.8 km^2 have been recorded, whilst the figure in Pembrokeshire is one pair per 0.4 - 0.8 km^2 (Lloyd, 1980). Most activity takes place at night, though in summer foxes may continue to be active after dawn. However, in urban areas foxes are usually only visible between midnight and dawn, an adaption to avoid humans.

The fox's adaptability extends to its choice of food, with rabbits, rodents, birds, invertebrates, fish and carrion all taken. Cannibalism of litter mates also occurs. Breeding takes place at ten months when four to five young are raised in March. Maximum life expectancy in the wild is ten to eleven years, but the high level of mortality means the survival of a pair from one year to the next is unlikely. In mid Wales, adult mortality is 57% (Harris and Lloyd, 1991).

Foxes are widespread in Carmarthenshire, both in traditional rural areas and in the urban south east. One of the earliest references to foxhunting in the county comes in Evans (1933), who recalls the Cryngae hound pack of Penboyr, home of Llywellyn ap Gwilym Fychan (uncle of the poet Dafydd ap Gwilym) which hunted in the 14th century. The fox was evidently abundant, for the hunt once apparently killed twelve foxes in three days. Numerous references to the species in vestry books, relating to the bounty paid on the species, show that foxes have probably always been numerous in the county. Carmarthenshire has been described as having ... *the reputation as being one of the most sporting counties in the Principality* (Morris and Parkinson, 1897).

Whilst its population may have declined during the Edwardian heyday of the gamekeeper, the fox's adaptability and its catholic diet have always ensured its survival. Foxes were ready to take advantage of any lightening of persecution. Matheson (1955) records the fox as surviving in large numbers, having multiplied greatly during the Second World War, an observation supported by local people who remember that period. They also saw them more regularly in the upper Tywi valley in the mid fifties, apparently because of having to hunt in more open areas following the loss of rabbits to myxomatosis. Local people connected with hunting suggest that the fox is now more common than it has ever been. Whilst hunts tend to exaggerate the number of their quarry to justify their own existence, there can be little doubt that the fox is very common in Carmarthenshire. However, despite claims of the farming community that hunting is essential to keep foxes under control, there is no evidence that hunting pressure has reduced fox numbers. Mortality caused by man probably removes foxes that would otherwise die in any event because of lack of available territories.

Despite the antagonism of sheep farmers, this attractive species remains a common feature of our countryside. In a country where most of our large predators were exterminated long ago, it is pleasing to note that at least one large carnivore has resisted eradication.

PINE MARTEN

Martes martes

BELE

Very rare, mainly recorded in the north and east.

The pine marten is one of Carmarthenshire's most elusive and controversial species. It is locally common in parts of Scotland, but considered very rare in England and Wales (Harris *et al.*, 1995). It is thought that, overall, the pine marten is extending its range (Arnold, 1993).

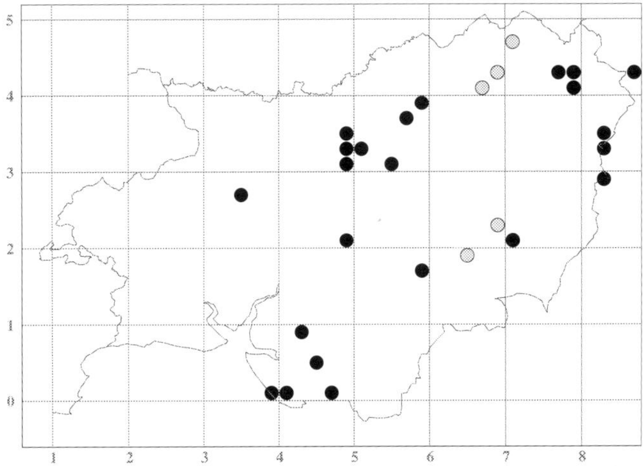

Pine martens are solitary animals, living in deciduous and conifer woodlands as well as more open country such as rocky and craggy areas. Even where habitat quality is good, the density of animals is low, with around one pine marten per 200 - 500 ha (Morris, 1993). They feed on rodents, birds, beetles and berries, most of the food being found on the ground, although pine martens are agile climbers. Breeding takes place in March or April when one litter of one to five young is raised. The main predator is man, with mortality caused by roads, trapping and shooting, despite the pine marten's fully protected status under Schedule 5 of the Wildlife and Countryside Act.

Once common throughout Britain, pine martens have declined, like many other predators, because of persecution. The species is vulnerable because of its territorial system, with one male defending one or two females, which results in low population densities. Breeding may be compromised when this fragile territorial system breaks down. Habitat fragmentation is a significant problem in pine marten conservation, and the species' low breeding rate makes it vulnerable to chance extinctions. Illegal persecution remains a major problem.

The history of the pine marten in Carmarthenshire has been reviewed by Morgan (1993), which details many of the early records for this species in the county. The first reference to the species is found in The Laws of Hywel Dda (see Chapter Two). The price of a marten's pelt was fixed at 24 pence (a great deal at that time), as opposed to the value of the wolf and otter at eight pence. The first modern reference in the county is found in Harting (1893) who recounts the story of Mr Gwynne Vaughan, his father-in-law, hunting and killing a pine marten with hounds near Cefn Farm between Llandovery and Llanwrtyd.

Barker (1905) notes the pine marten as having become extinct, the last being shot near Kidwelly 'some years ago'. This may refer to two individuals, reportedly shot by Mr Thompson of Glyn Abbey in 1880. There are also two unsubstantiated records from the Pendine area, though these should be treated with caution. Langley and Yalden (1977) also suggested that the pine marten had probably become extinct in Carmarthenshire around 1870, although they considered it to be still present in nearby Brecknock.

However, the demise of the pine marten had been greatly exaggerated. Matheson (1955) records unconfirmed reports from Carmarthenshire in the mid 1950s, helped by the spread of conifer plantations in mid Wales. In 1971 an animal unmistakeably a marten was run over between Cwrt y Cadno and Rhandirmwyn (Panting, 1972), and two other sightings were reported from the same area in the mid 1970s. Two forestry workers who had experience of the species from Scotland saw another in Brechfa Forest at Trawscoed in 1977. In 1982 a decayed corpse which was subsequently confirmed as a marten was found near Llandeilo. Two years later, Chris Mead, of the British Trust for Ornithology, saw an individual 'go up one tree, down it, across to the next mature tree and up that' near Rhandirmwyn. A nearby raptor's nest failed shortly after this and the marten was thought to be involved.

Two systematic surveys, in the 1980s and 1990s have painted a pessimistic view of the future of the pine marten, not only in Carmarthenshire but in the whole of Wales. Velander (1983) found no evidence of martens in Wales, though she acknowledged that this could be a result of the difficulty of locating scats when populations are at low density. McDonald *et al.* (1994) carried out a survey for martens in north and mid Wales, again using scats as signs of presence. Pine marten scats have a characteristic shape and smell, being twisted with a sweet odour. They found only three scats which they could positively ascribe to pine martens, including one at Rhos Wen in Brechfa Forest. They concluded that no viable population of pine martens is present in Wales, although a few individuals may still persist, and that the species was, at best, on the brink of extinction in the Principality.

Despite the orthodox view that the pine marten had virtually disappeared from most of Wales, including Carmarthenshire, sightings continued to be reported. In 1995, the Vincent

Wildlife Trust began a research project throughout Wales to actively seek out records using a poster campaign, articles in the local and national press and items on radio and television. Up to 1997, this had produced 83 reports thought to be authentic from throughout Wales. Sightings were graded on a scale from 1 to 10, where 1 indicated that a misidentification had certainly occurred and 10 denoted a sighting that was without doubt a pine marten.

All records scoring six or better (considered to be at least 'probable' pine martens) have been included on the distribution map (Messenger *et al.*, 1997). Reports are non randomly distributed, clumped in the north of the county around the Brechfa area and the upper Tywi valley, with others near Pembrey. Those in the north are of particular interest as they are part of a larger aggregation of records that skirt the northern edge of the Brecon Beacons, in a broad arc across southern Powys. Whilst some of these records may still be erroneous, can all these independent observers have made mistakes?

Most of the initial effort in pine marten recording was expended in the large conifer forests in north Carmarthenshire. This was done because it was believed at the time that these areas offered the best habitat. It is in such areas that it occurs in Scotland and even the name pine marten hints at some association with conifers. However, pine martens are far from exclusively found in pines. This is certainly the case in Ireland, where they occur in open country and in landscapes not unlike much of Carmarthenshire. It is likely that pine martens actually require a mosaic of rough unimproved land, over which to hunt, and woodland of some kind in which this agile arboreal species can shelter. The forests of Brechfa, Crychan and the upper Tywi valley all have adjacent suitable open country such as the heather clad Mynydd Llanllwni and the wild expanse of Mynydd Mallaen. River valleys may be of particular significance as routes by which martens move around the countryside.

The true status of this, one of our most attractive species, is still unclear, and the animal itself remains elusive. However, given the number of reported sightings, clustered as they are in areas of apparently suitable habitat, the evidence that a relict pine marten population still exists in Carmarthenshire seems compelling. Its long term prospects remain uncertain.

STOAT

Mustela ermina

CARLWM

A common, but under recorded, species in most habitats.

The stoat is a common species, but is possibly declining (Harris *et al.*, 1995). Records are obscured by problems in distinguishing this species from the weasel.

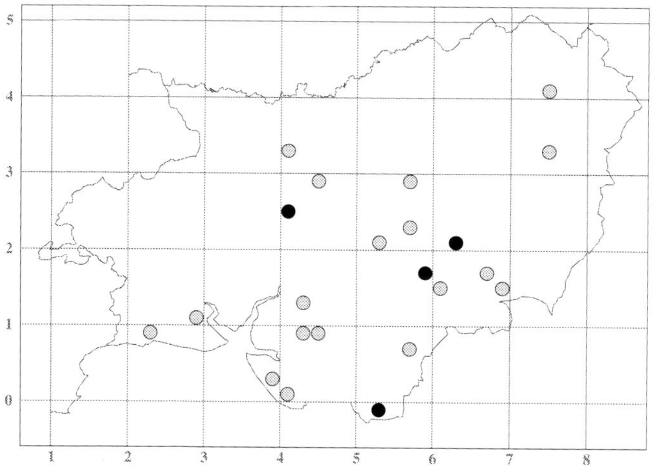

Stoats are found throughout Britain in any habitat which offers cover and prey, such as farmland, woodland, moors, marshes and mountains. Their diet consists of rabbits, water voles, brown rats, grey squirrels and occasionally earthworms. Stoats can continue to kill and cache prey irrespective of satiation when prey is plentiful. This is not, as many have supposed, an indication of 'blood lust' by the stoat, but simply a predator's reaction to an abundance of prey in an unpredictable environment. When an animal kills dozens of hens in one building it does so because its natural behaviour patterns have been 'overloaded'. Its actions should not be interpreted anthropomorphically, however distressing they may appear to the human observer.

Breeding takes place in spring when six to twelve young are raised in a single litter, with the number of young raised influenced by the availability of food. However, mortality can be high with an annual turnover of 40 - 78%. Shortage of food is the main cause of death, though owls occasionally take stoats (King, 1991a).

The stoat has probably always been common in the county, with unimproved pastures and woodland providing abundant prey. Barker (1905) regarded it as common, whilst Brunker (*c.* 1950) notes it as often seen, and possibly on the increase. Matheson (1955) stated that the stoat was often caught in rabbit traps, so much so that in places in south west Wales, it was locally, if temporarily, exterminated. Myxomatosis was also thought to have had some affect on the stoat, but quantitative information is lacking. In any event, it is likely that the species would have survived by switching to alternative prey. Conversely, Davis (1955) reported that the stoat was becoming more numerous in north Carmarthenshire. Stoats were also noted as occasional on Laugharne-Pendine Burrows (Thomas, 1955).

In a short description of the mammals around Llanelli 'Deerfoot' (1929) gives a graphic, if emotive, description of a hunting stoat ... *relentlessly he pursues his prey, never faltering in his rapidly gaining stride. Master Bunny running against overwhelming odds, screams piteously and surrenders to the merciless fangs of the stoat, whose black list is ever being added to by crimes of this sort.* Such anthropomorphic views of our carnivores have sadly ensured their continued persecution.

Stoats remain relatively common in Carmarthenshire, though it has probably declined, particularly in the west where habitat loss has been greatest. However, records are few compared with other mammal species, reflecting the difficulties of observing and identifying mustelids in the wild. Their small size also makes them inconspicuous as road casualties, the main source of records for the larger polecat. Slightly fewer stoats than weasels were trapped in recent years by gamekeepers in the Llandeilo area, with less than 20 caught per year.

WEASEL

Mustela nivalis

GWENCI

Common, possibly more numerous than the stoat.

Weasels are common throughout Britain. There is a peak in sightings nationally during spring, coinciding with increased activity when mating (Arnold, 1993).

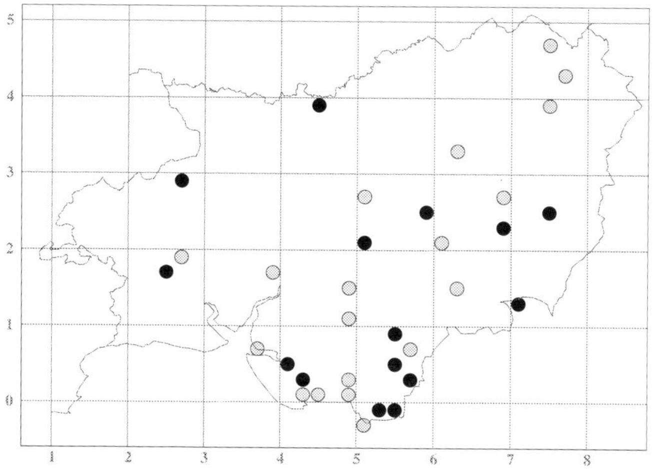

The weasel occurs sympatrically with the stoat across the northern hemisphere in a wide range of habitats, but is less common where small rodents are scarce. The two sexes occupy separate territories, with male territories generally larger than female. Rodents are the primary food source with field voles, bank voles, and wood mice all taken, as well as baby rabbits. One or two litters are raised each year, with four to six young in each. Weasel populations are highly unstable, reflecting prey availability, and life expectancy is about one year, little different from their prey. Weasels are themselves preyed upon by hawks, owls, foxes, cats and mink (King, 1991b).

Like the stoat, the weasel has been regarded as common in Carmarthenshire, although records are infrequent. Barker (1905) noted it as common and Brunker (*c.* 1950) regarded it as the commonest mustelid, often being caught in mole traps. The suggestion that weasels were more common than stoats was questioned by Matheson (1955), who presented contradictory data from elsewhere in Wales on the relative proportions of the two species. Quantitative data on the species in the county is lacking. Most reports in *Nature in Wales* refer to anecdotal records of little value, such as the two animals seen crossing roads in the county in 1964; one at Llandeilo and the other near Tavernspite (Anon., 1964). Local gamekeepers report killing about 30 per year in the Llandeilo area in recent years.

Records are infrequent for the same reasons as those documented for the stoat. However, given their widely dispersed nature, there can be little doubt that the weasel is a common inhabitant of the county where suitable habitat occurs. Whether the weasel is genuinely more common in the county than the stoat is uncertain. Local naturalists attest that they see the former species far more frequently than the latter, whilst gamekeepers report catching the weasel marginally more frequently than the stoat. In a school exercise in the Rhandirmwyn area during the 1960s, children brought in 12 weasels for every stoat. However, Harris *et al.* (1995) were of the opinion that, throughout Wales, stoats marginally outnumbered weasels.

The future of this diminutive carnivore, like so many species, depends largely on habitat conservation in the wider countryside, and the continued survival of prey rich habitats such as woodland, unimproved pastures and hedgerows.

POLECAT

Mustela putorius

FFWLBART

Fairly common, occurring throughout the county.

Polecats are common within their restricted British range of Wales and the border counties, though isolated individuals have been noted as far as the east midlands. Most are seen in April or September, coinciding with sexual activity in the spring and dispersal by juvenile animals in the autumn (Arnold, 1993).

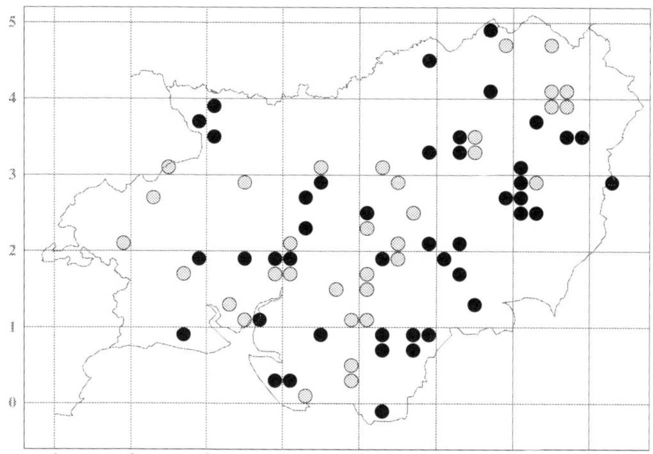

Polecats occur in a wide range of habitats including woodland, forestry plantations, marshes, river banks, sea cliffs, sand dunes, farms and even urban areas. Scat analysis has shown that rabbits and hares make up the largest component of the diet (37%), followed by small rodents (32%), birds (19%) and amphibians (6%) (Blandford, 1986). Carrion is also taken.

Breeding takes place once a year in May or June, but juvenile mortality is high. Life expectancy at birth is eight months, although polecats have been known to live up to five years in the wild. Man is the chief predator, with 86% road kills recorded in one study (Walton, 1976).

The polecat's reputation for bloodthirstiness resulted in it being declared vermin under the statutes of Elizabeth I, and persecution of the species reached its zenith at the end of the 19th century (Birks, 1993). Until the middle of the 19th century, the polecat was widespread and common in Carmarthenshire. However, the effect of gamekeepering and 'vermin' control in the county was such that by about 1880, it could only be described as widely distributed in the north of the county (Langley and Yalden, 1977). At the turn of the century it was effectively extinct in the south of Carmarthenshire (Millais, 1904), but still sporadic in the north. Barker (1905) regarded the polecat as still fairly common around the Carmarthen area. By 1915, the polecat was confined to a heartland around Aberystwyth along with that other refugee from Edwardian persecution, the red kite, and was only sporadic in the north of the county. The species therefore appears to have only narrowly escaped extinction in the country. The First World War, and the departure of many gamekeepers for France, came just in time for the polecat and, freed of persecution, the population began to rise again. By the early 1930s, the polecat had recovered much of its lost ground. Matheson (1932) described it as 'fairly common' in parts of Carmarthenshire, along with the adjacent counties of Ceredigion and Breconshire, suggesting the north of the county was still the stronghold at that time.

The polecat continued to recolonise the county over the next half century. Brunker (*c.* 1950) recorded it from six localities in the Llanegwad area and regarded it as increasing in number, and being darker coloured than those in Ceredigion. Local people report that the polecat was quite common in the Llandeilo area by the 1960s and was regularly trapped on nearby estates. By the 1970s it was regarded as commonplace (Panting, 1971), with records around that time from Cilycwm and the Carmarthen area. However, Lockley (1970) noted it as rare in the Talley area, giving its distribution in Carmarthenshire as being north of a line from Llangadog to Newcastle Emlyn, and only common north of Pumpsaint and Llanybydder.

Today the polecat is widely distributed throughout the county, and can be encountered almost anywhere, even in urban Llanelli. Sightings of this elusive species are rare, however, which made an observation of a female crossing a road with ten young near Newcastle Emlyn in May 1995 all the more remarkable. It is likely that its presence would go largely undetected were it not for the number of road casualties which can be seen throughout the county annually. The polecat is now secure in Carmarthenshire, a sound base from which it has made a welcome return to south east Wales and, more recently, parts of England.

MINK
Mustela vison
MINC

Reasonably common on most river systems.

The mink is an introduced species originally native to North America. First imported in 1929, it is now widespread in mainland Britain.

Mink occupy a wide range of aquatic habitats, but can occur away from water in places where prey, for instance rabbits, are abundant. They favour eutrophic waters, especially if there is abundant emergent vegetation, but they are also found on coasts. Due to the nature of the preferred habitat, ranges tend to be linear, between one and six kilometres.

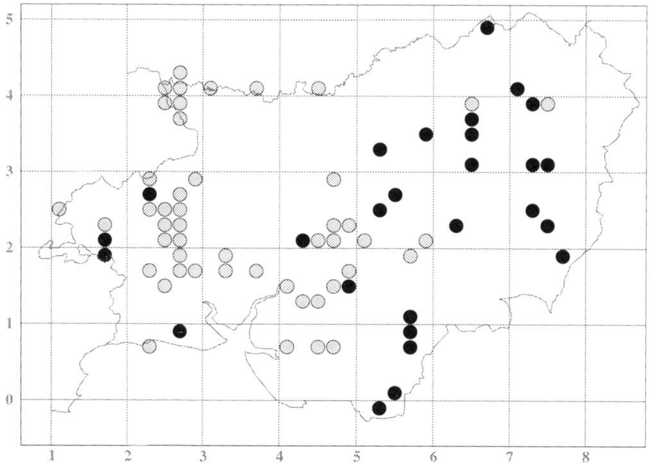

The mink is very much an opportunist, with mammals, fish, invertebrates and birds taken in proportion to their availability. Detailed studies suggest that, contrary to popular belief, mink have little effect on prey overall, though some reduction can occur in habitats that are already sub-optimal. Breeding takes place in May when a single litter of four to six young is raised. Mortality factors are not well known. Large numbers are shot by man, but it is the mink's own territorial system that probably exerts most control over the population (Birks and Dunstone, 1991).

Mink had been imported into Britain for the fur trade from 1929 and, although escapes occurred, breeding was not proved in the wild until 1956, in Devon (Harris *et al.*, 1995). At that time the species was unknown in Wales, in the wild at least; it does not even merit a mention in Matheson (1955). There are several versions of the origins of mink in west Wales. Holiday (1961) claimed that animals had probably escaped from unofficial 'back garden' breeders in 1958. Another story holds that mink were apparently deliberately released from a farm near the coal yard in St. Dogmael's, Pembrokeshire, either by the owner or by a vindictive employee. Whatever the source, mink were certainly established on the Teifi, Cych and Taf, as well as across the border on the Eastern Cleddau in Pembrokeshire by the late 1950s and early 1960s (Dunstone, 1993). Thompson (1964) notes that the mink had been spreading along the tributaries of the Teifi on the Carmarthenshire border ... *for the last six years* ..., and describes a cache of 21 dead trout found in the area believed to have been killed by mink. Lever (1977) reported the mink to be established on the Teifi and the Taf by the early 1960s and Hill (1964) records the species as thriving on the Teifi at about the same time.

An extensive trapping campaign organised by the Welsh Office Agricultural Department in west Carmarthenshire had little effect on the spread of the species. By 1971 mink were well established on the Taf, Teifi and the lower reaches of the Tywi (Corbet, 1971). Local gamekeepers report that mink first became established on the Tywi in the late 1960s. Since 1970, mink have moved eastwards in the county, to meet another population centre moving westwards from the English border and are now present on every Carmarthenshire river. In the late 1980s, a concerted effort at trapping by one person caught approximately 600 in one year in the Dryslwyn area (mink apparently being very easy to catch) but this is likely to have had, at best, only a temporary brake on expansion. Populations are also established in the industrial south east of the county, with animals having been seen at Glynea Pond, Bynea, and are regularly recorded at the Wildfowl and Wetlands Trust Centre at Penclacwydd. A small number of completely black animals have been recorded, though most are of brownish-black colour.

Much has been written, some quite ill informed, on the potential impact of the mink on British wildlife. The species itself retains a fearsome reputation with local people, being held responsible for the death of everything from sheep to moorhen chicks. However, in Carmarthenshire it exists alongside the polecat, and the recovery of the otter in the last ten years, when mink were already well established, suggests little competition between the two. There has certainly been some impact on the water vole elsewhere in Britain, causing a fragmentation of the population, but this seems worst in places where the habitat is already sub-optimal. More sympathetic management of our rivers as living systems rather than as drains to transport water to the sea would give the water vole a better chance of withstanding mink predation.

In any event, the mink is here to stay and our riverine management should take the species into account. It is worth remembering that it is only an accident of glacial history that prevented the European mink *Mustela lutreola* from colonising Britain. It would therefore appear that the mink has exploited that vacant niche. No doubt mink will continue to be a controversial addition to our fauna.

BADGER

Meles meles

MOCHYN DAEAR

Ubiquitous throughout the county, even in urban situations.

Badgers are considered common in Britain (Arnold, 1993), though they may be locally scarce because of persecution.

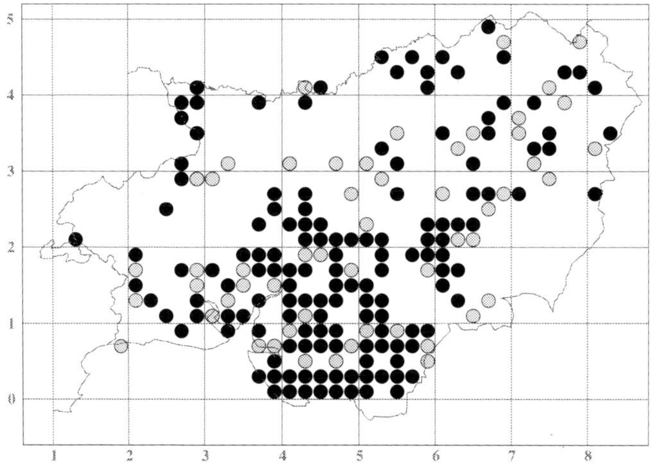

Badgers are probably more numerous now than in 1900. Persecution was severe during the last century and continues to the present day, with 10.5% of setts throughout Britain showing signs of digging activity by humans (Cresswell *et al.*, 1990). Up to nine thousand setts continue to be dug each year (Morris, 1993), and habitat fragmentation may also affect badger populations.

Badgers can occupy a range of habitats including woodland, scrub, hedges, quarries, sea cliffs, moorland, open fields and urban situations, though they prefer woodland. Badgers live in social groups of two to twenty-three animals, with a territory of 30 ha in the best habitat, but this may increase to 150 ha in more marginal areas. Earthworms are the most important food, though they also take insects, small mammals, carrion, cereals, fruit, birds, amphibians, reptiles, molluscs and fungi. Between one and four young are born in February, and emerge from the sett for the first time in April. Cub mortality is high, with 50 to 65% dying in the first year. Few badgers exceed six years in the wild although they have no natural enemies in Britain. Most deaths are human related, with roads being a major factor (Neal, 1986).

It appears the badger may always have been a common resident of Carmarthenshire despite being subject to persecution during the last century. Pease (1898) regarded badgers as reasonably plentiful in mid and south Wales and Barker (1905) noted the badger as ... *still fairly common* ... near Carmarthen, perhaps hinting at the level of persecution at that time. 'Deerfoot' (1929b) also notes the badger as being ... *hunted down as the enemy of the farmer* ... in the Llanelli area. Perhaps surprisingly, Brunker (*c.* 1950) does not mention the badger, though this must be an oversight as it is scarcely credible that the species was absent from the parish of Llanegwad at that time.

Matheson (1955) notes the badger as being common in most parts of Wales apart from the highest hills, and notes an erythristic (i.e. reddish) individual run over by a car in the county in 1953. Both albino and erythristic individuals were noted in the Alltwalis area in 1970 (Panting, 1971). Erythristic individuals were also recorded at Trap and Rhydcymerau, the skins of which were sent to the National Museum of Wales, and pale individuals were noted near Tywyn Point in the late 1970s (R. Pryce, pers. comm.). Neal (1986) published a map of badger abundance suggesting that west Wales, including north and west Carmarthenshire was a stronghold for the species in Britain, although this map was mainly the result of guess work. Badgers were also 'frequent' in the Llanelli area and only considered infrequent or absent from the Mynydd Du area, presumably because of the absence of suitable terrain for digging.

Today, despite continuing persecution, the badger is still a widespread and common species in Carmarthenshire, even on the edge of urban Llanelli. In general, the steep wooded valleys that abound in Carmarthenshire are the preferred habitat. However, setts are found in a number of locations, including open grassland and derelict colliery sites, where coal spoil provides an ideal medium for digging setts. Eight setts are also present on sand dunes at the Ministry of Defence establishment at Pendine Burrows. Setts may be extremely long lived, with at least one known to have been in continuous occupancy for 50 years. Badgers are also present on most of the farms in the Dinefwr area that are currently taking part in the 'Tir Cymen' countryside stewardship scheme. The distribution map for badger in Carmarthenshire, which includes sett records, as well as sightings of individual animals, road casualties and tracks and signs, is biassed towards the east and south. This is largely because of the number of local naturalists in this area and their readiness to make records available. In west Carmarthenshire, a recent split in the Badger Group in that area has made access to information difficult. However, badger enthusiasts in the west maintain that the species is present in most woodlands, as in Dinefwr. A complete map of badger distribution would leave few blank tetrads.

Although badger baiting represents no threat to the survival of the species in the county, at least at its current intensity, it is a serious animal welfare issue. Most baiters come from outside the county, with a concentration in the Swansea, Neath, Port Talbot and Rhondda areas. Approximately 150 people may be involved, with at least one person being known to have killed 20 to 30 animals in one year. Maps confiscated from diggers show that they probably have a more comprehensive knowledge of the location of setts in the area than the nature conservation bodies. Digging may occur with the connivance of the landowners, especially if they wish to have the badgers removed, and in such circumstances, they will exterminate whole setts. However, when the landowner is not aware of the activity and the diggers require more badgers for future 'sport', they may take only one or two animals.

Most badger baiting occurs between February and May when young are present, and when the badgers are therefore more likely to put up a good fight. Usually a terrier is put into a sett to corner the animal so that it can be uncovered by digging. Once the badger is captured, it may be baited at the sett in a makeshift pit or taken away. To give the dogs a better chance the badger may have its back legs broken or be blinded, and bets will be made on whose dog will despatch the animal. Finally, if the badger is not killed, it will be shot or hacked to death with a spade and the body often dumped on the road to disguise it as a traffic accident. One can only wonder at the mentality of people who are capable of such brutality.

West Carmarthenshire, along with the south west of England is also a contentious area for the connection between bovine TB and badgers. The Ministry of Agriculture regularly investigates badger setts in the west of the county. Although it is undoubtedly true that badgers carry bovine TB, the limited amount of research done does not suggest a simple direct link between badgers and the disease. There are, for example, areas of Britain where badgers are common but bovine TB is absent. There is some circumstantial evidence that badgers can transmit bovine TB to cattle, but the mechanism is unknown. Research by the Ministry of Agriculture suggests there is no link between badger density and cattle infection (Pain, 1997). More research is required to answer the doubts about this most controversial of issues.

Despite all the odds, the badger remains a common animal in the county. The Wildlife Trust: West Wales regularly holds badger watching evenings on its reserves, such as Castle Woods near Llandeilo, and the experience of seeing a badger only a few metres away as it snuffles for peanuts is an unforgettable one, enjoyed by residents and visitors alike. It is to be hoped that this, the ultimate survivor, will endure to delight future generations in years to come.

Badger protection is the subject of complex law, but was consolidated in the Protection of Badgers Act 1992. This made it an offence to damage a sett or to disturb a badger whilst it is occupying a sett, as well as giving protection to the animal itself.

OTTER

Lutra lutra

DYFRGI

Formerly scarce, now present on most river systems.

The otter is locally distributed in Britain and is considered endangered nationally (Arnold, 1993).

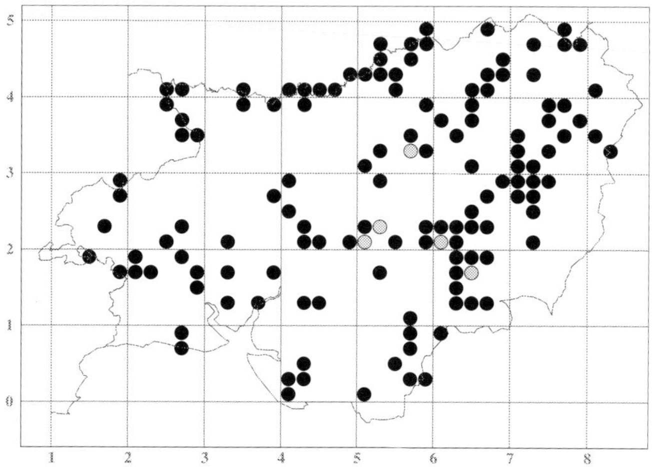

Otters inhabit many aquatic habitats including lakes, rivers, streams, marshes and coastal areas, although they can travel some distance overland, particularly when dispersing between river catchments. Availability of lying up and breeding sites is critical to determining the otter population of a particular water body. Food is mainly fish, but crayfish, frogs waterfowl, water voles and even rabbits may very occasionally form part of the diet.

Between one and five young are raised in a litter, breeding activity taking place throughout the year. Otters become sexually mature in their second year. They have few natural predators, with starvation being the main natural killer and road casualties also being a major cause of mortality (Chanin, 1985).

Like other carnivores, otters endured persecution until recently as 'vermin'. From 1930, hunting killed 400 animals annually (Morris, 1993). Paradoxically, supporters defended this on the grounds that it was beneficial to otters! Otters were still fairly common in 1950, but a reduction in hunting success from 1960 indicated that the species was in decline, and by 1970 otters were rare or absent in most parts of Wales. Hunting did not cause the decline however, since populations did not rise once the practice was made illegal. Mink have also been blamed for the decline of the otter. However, otters and mink exploit different food resources and, in any event, the otter has recovered in recent times with mink already firmly established on many rivers. As in other similar situations, habitat loss was the main issue. In the otter's case, the decline was caused by development, drainage and unsympathetic river management. Pesticides may also have been a factor, causing impaired breeding success. Pollution of rivers such as the Aman, Gwendraeth and Loughor by leachate from coal mines may have been a contributory factor in those areas.

In the last century, despite a continuing level of persecution, otters appear to have been relatively common in most of the county. Davies (1858) notes the presence of otters on the Tywi saying that ... *salmon are sometimes found lying wounded and exhausted from loss of blood ... attacked by ... otters, more for amusement than food.* Hunting was a popular pastime and the *Llanelli Guardian* reported a four foot otter being killed on the River Cennen in 1879, the same paper recording 'a fair days sport' by Captain Glasevin's otterhounds near Llandeilo in 1886. The extensive marshes that formerly existed between Pembrey and Kidwelly were also the haunt of otters and, as the *Llanelli Guardian* explained in 1883, otters did not always come off second best in confrontations with humans ...

> *CHASED BY OTTERS. Between Pembrey and the Gwendraeth river there is a vast tract of land known as Pinged marsh ... which forms the haunt of a prolific number of otters. About dusk last Thursday a lad 12 years old ... was returning over the Burry Port railway line and while near Tymawr crossing, at which there is a deep culvert running under the line, he saw three otters, one large and two smaller ones. The three at once made for him and it was only after a hard race that he escaped injury and got shelter in a cottage not very far off, whence he got company to escort him home, the company being well equipped for the water dogs who however had escaped to their den* (Anon., 1883).

The otter appears to have remained reasonably common, despite the attentions of hunters, until shortly after 1950. Barker (1905) notes the otter as common near Carmarthen and 'Deerfoot' (1929b) recorded that otters could be seen in the vicinity of Swiss Valley and other rural areas around Llanelli. George King (1930), then the superintendent of the Towy Fishery Board, sent a number of replies in response to a request for help from the University of London Animal Welfare Society, who were undertaking a survey of the species at that time. He considered the otter numerous or even abundant throughout the county except in rivers polluted by industrial effluent, and he records that the otter was ... *so numerous in recent years that certain riparian owners ... have taken measures to have them destroyed ...*, and that ... *coracle netsmen ... frequently commented about them, generally in an unkind manner.* The otter was also blamed for causing economic damage as a result of salmon and trout predation, the cost of an otter kill being ... *no less a sum than £7 10s.*

Carmarthenshire was the province of two major otter hunts; the Hawkstone, whose country was the Teifi and the Cothi, and which quartered their hounds at Alltyrodyn when in the county and, the Pembrokeshire and Carmarthen which operated on the Tywi and in west Carmarthenshire, as well as the Cleddau in Pembrokeshire. The Ponsonby Lewis Hounds also hunted the Tywi and Teifi, at the invitation of the Pembrokeshire and Carmarthen. In the 1950s, otters were considered plentiful, with the hunts apparently able to pick up the scent of an otter within 100 m of the release point on any river. Details of the best areas to hunt were passed to the hunts by the river bailiffs who, as a reward, would be paid ten shillings if their information lead to a kill. Occasionally, otters were so numerous that more than one could be killed in a day. Stripped of its glamour, however, otter hunting was a ruthless business. Occasionally otters would be brought to ground with bleeding feet having been chased over long distances. Otters themselves also occasionally managed to kill the hounds by drowning them, as reportedly happened near Abergorlech in the 1950s. After a successful hunt, new recruits to the hunt would be 'blooded', by smearing blood on the face, as still happens in fox hunts today and, if the animal was a male, they would remove the penis as a trophy.

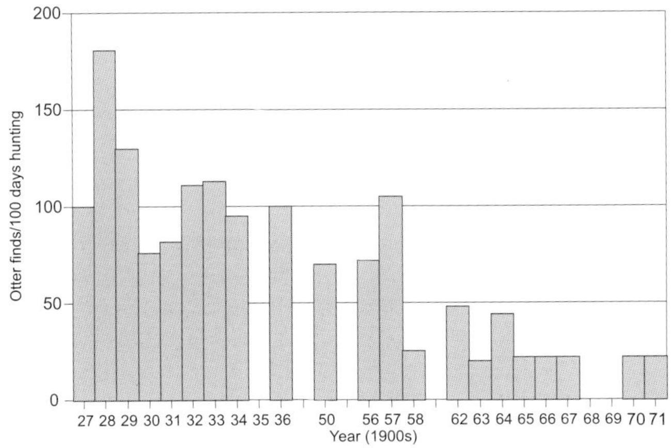

Graph 7 - Pembrokeshire and Carmarthen Otter Hunt Results 1927 - 1971

Data from the Year Books of the Pembrokeshire and Carmarthen O.H. (Unpublished manuscript held at Carmarthenshire Archive Office ref. Buckley 32 and 42). No data is available for the following years: 1935, 1937 - 1949, 1951 - 1955, 1959, 1960, 1968 and 1969.

Graph 7 shows the number of otter 'finds', i.e. the number killed plus those located but which escaped, adjusted for the number of days hunted; generally the hunt was out for between 30 and 40 days a year. Numbers caught were broadly stable until 1958 when there was an abrupt decline. That year's results were attributed to bad weather, but numbers failed to recover and continued to decline for the next 13 years until otter hunting ceased on the river.

To conservationists unable to remember the situation before the mid 1960s, it is hard even to conceive of the hunting of such a rare species. Some of the notes in the Pembrokeshire and Carmarthen Hunt year book give a flavour of the times:

1932 -	-	*'a good supply of otters in Carmarthenshire.'*
3 September 1962	-	*'a splendid 6 hour hunt eventually killing a 13 lb bitch.'*
9 September 1963	-	*'a first class hunt with a 25 lb dog otter killed below Glanrhydw.'*
1965 -	-	*'a deplorable season.'*
1966 -	-	*'a very poor season, otters scarcer than ever.'*

By the late 1960s only one otter was being killed a year and in 1971 only four otters could be found, the hunt failing to catch a single one, compared with the heyday of the hunt in the late 1920s when up to 20 were killed a year and 40 were found but subsequently escaped the dogs. Otter hunting finally became illegal in 1978 when the otter was added to Schedule 1 of the Conservation of Wild Creatures and Wild Plants Act, 1975, although the paucity of animals meant that hunting had ceased some years before. The tide of public opinion had also turned against hunting, for there were anti hunt demonstrations against the Hawkstone in the Brechfa area in the 1970s.

The decline and subsequent revival in the otter population in Carmarthenshire has been one of the most notable trends in mammal populations in the last 30 years. In contrast to all other mammal species in Carmarthenshire, there is high quality systematic survey information on the status of the otter in the county from the late 1950s to the present day (Andrews and Crawford, 1986; Andrews *et al.*, 1993; Crawford *et al., 1979;* and Stephens, 1957). These surveys are based on river catchments, rather than counties, which in Carmarthenshire are divided into the three following areas; Teifi and tributaries, Tywi and Taf and associated tributaries and finally the Loughor and tributaries. Surveys were carried out by looking for the characteristic droppings, or 'spraints' at selected spots, usually near bridges. Most of the following account of the recovery of the otter is based on these sources and consequently also divided by river areas.

In the late 1950s the Teifi was described as excellent for otters, with animals present along the whole length. Despite the decline of the 1960s and 1970s otters were still found all along the river where it forms the Carmarthenshire border. The Tywi was also well stocked, with otters also present on the Cothi and Gwili, and were described as 'numerous' on the Gwendraeth Fach. By contrast there were no signs on the Loughor catchment. However, one local person remembers seeing the species on this river near Penybanc in 1953 or 1954 and recalled seeing a large male die, possibly of poisoning, in 1955 or 1956. Signs of otters were found on the Cothi in 1973 and there is an unconfirmed report of an otter being killed on the Taf by the Pembrokeshire and Carmarthen Otter Hunt in 1976.

Following the survey of 1977-78, the Teifi was still adjudged to be one of the best rivers for otters in Wales, with signs along the whole of its Carmarthenshire length. Otters were also recorded on the Cych and Nant Bargod, but absent from the Tyweli and Nant Arberth. Signs were considered sporadic on the Tywi, with only 14% of sample sites being positive. Otters were still present on the Cothi, but were absent from the Gwendraeth Fawr and Fach and the Afon Gwili. The Tywi, along with the Gwydderig and Dulais, had only a few positive sites. There had therefore undoubtedly been a decline on this river system in the previous twenty years. Otters were still absent from the Loughor. It appears that the otter reached its nadir in Carmarthenshire at about this time.

The survey was repeated in 1984-85, but unfortunately comparisons with the first survey some eight years earlier was complicated by differences in weather conditions and survey technique. Otters appeared to have remained stable in the Teifi catchment, with good results on the main river and the Cych, and animals also appearing on one additional tributary, the Tyweli. The Tywi showed the biggest increase of any river in Wales, from 14% positive sites to 68%, although this was also influenced by survey methods and weather. Tributaries of the Tywi also scored well, with the Gwili, Gwydderig and Cothi all giving good results. The Taf and its tributaries the Cynin and Cywyn also had a good result, and otters appeared on the Gwendraeth Fach, although the Gwendraeth Fawr remained negative, as did the Loughor system.

The most recent survey was in 1991. Otters continued to be present on the Teifi, with most main river sites proving positive, together with those tributaries where otters had been recorded in previous years. A new positive result was noted on one tributary, the Nant Arberth. Signs were also present on the Tywi from Llyn Brianne to the sea. Otters expanded on the Cothi, to the Twrch, Melinddwr and Gorlech, and were also present throughout the Taf system and associated tributaries. The Gwendraeth Fawr was negative during the survey, but a subsequent field by field survey confirmed otters on this river. Most excitingly of all, otters reappeared on the Loughor, with positive signs on the main river along with the Morlais, Gwili, Marlais and Aman. Only the Afon Lliedi was totally negative. However, otters have been recorded on the Lliedi since that survey.

Thus, from a low point in the mid 1970s, the otter had returned to most of its former haunts in Carmarthenshire. Today it is present on all major rivers, although numbers remain low in the industrial south east compared with the rest of the county. However, it is of particular interest that an otter was seen in Old Castle Pond in central Llanelli in 1995 (Lucas, 1996). Wildlife in urban areas in England, such as badgers and foxes, can receive national television coverage. That otters in Carmarthenshire's largest urban area can go largely unnoticed is an example of how the wildlife of Carmarthenshire is under valued by people both within and without the county. At least this individual received a better reception than an otter seen in the High Street in 1905, which was apparently chased through the town before being clubbed to death by the local inhabitants.

This good news of the otter's revival in Carmarthenshire is tempered by the fact that there are fewer otters on Carmarthenshire rivers compared with the Wye and the Severn, the best sites in Wales. Wales as a whole has fewer otters than Ireland or Scotland. The otter is doing well in Carmarthenshire, but it undoubtedly could do better. There is still much to be done on aspects of river water quality and habitat creation to secure the otter's future in the county.

Nevertheless, it is now relatively easy for the determined and patient naturalist to see this, one of our most charming mammals. The would be otter watcher is recommended to read Neil Mathew's evocative account of his observations on two otter families on the Tywi near Llandeilo in 1991 (Mathew, 1991a). However, most of us can only dream of having as close an encounter as he had one May morning:

> *Crouching ... I heard a couple of splashes out in the river ... Suddenly there was a ripple of water ... and a little head popped up above the surface, about two feet from my own. We looked at each other ... in mutual astonishment. The Otter dived and retreated a bit, soon coming up for another look ... I was addressed with a couple of 'Hahs' before it dived out again into the river.*

Otters are now fully protected under schedule 5 of the Wildlife and Countryside Act 1981.

GREY SEAL

Halichoerus grypus

MORLO LLWYD

Occasional off shore or in estuaries.

Grey seals are widely distributed on the west coast of the British Isles, with approximately 50% of the world population breeding in British waters (Arnold, 1993).

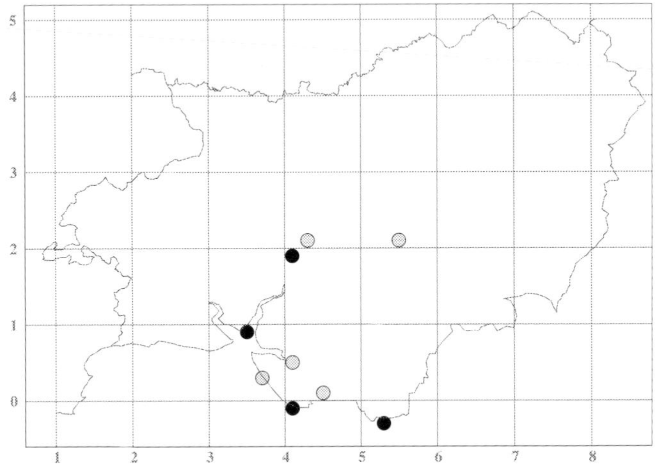

In south west Wales, the female grey seals raise their single pup on exposed rocky coasts, from September to December, on the coastline between Cardigan and Pembroke. The young seals become sexually mature at three to five years, though males are unlikely to be successful until they are ten years old, because of the physical demands of defending females. The maximum age is thought to be about 46, though few live beyond 35.

Man is the only significant predator, though they may be taken by killer whales and great white sharks. Respiratory and digestive tract disorders kill many adults, whilst young seals are often victims of starvation. However, little is known of grey seal biology away from the breeding sites (Anderson, 1991).

Grey seals do not breed in Carmarthenshire, despite the small amount of suitable habitat in the far south west of the county, between Amroth and Pendine. As this area is completely open to the public, it is too disturbed to be attractive. Most breeding in Pembrokeshire occurs on inaccessible coves and islands. Here they are safe from human persecution or the well intentioned but ill informed attentions of 'animal welfare' enthusiasts who 'rescue' pups left by females whilst the mother feeds in the open sea. The closest breeding population to Carmarthenshire is on Caldy Island.

Grey seals are occasionally seen inshore off Carmarthenshire coasts. The *South Wales Press*, under 'Burry Port News' on 12 October 1882 reported that ... *a seal measuring four feet was shot near the pier on Thursday last. Its colour was dark grey with black spots and weighed about 1 cwt. Seals are frequently observed in this channel but seldom caught.* Naturalists continue to report occasional grey seals offshore when watching seabirds, and other individuals have been noted off North Dock, Llanelli and elsewhere in the Burry Inlet. Grey seals are also regularly seen in the Tywi estuary, off Salmon Scar.

Readers may be confused by apparently 'inland' records. These refer to animals seen upstream in the Afon Tywi as far as Dryslwyn. Here they are unwelcome visitors because of their predatory habits on fish, and there are rumours that grey seals found in the river have been shot as a consequence.

WALRUS

Odobenus rosmarus

MORLO YSGITHROG

A Vagrant. One record.

Walruses are animals of the Arctic, living on pack ice, where they feed on clams and occasionally other seals. They occur rarely around British waters.

There is a single Carmarthenshire record, of one individual found dead on Cefn Sidan beach, 21 November 1986.

FALLOW DEER

Dama dama

DANAS

Common near former deer parks in Dinefwr. Escaped population at Pendine Burrows.

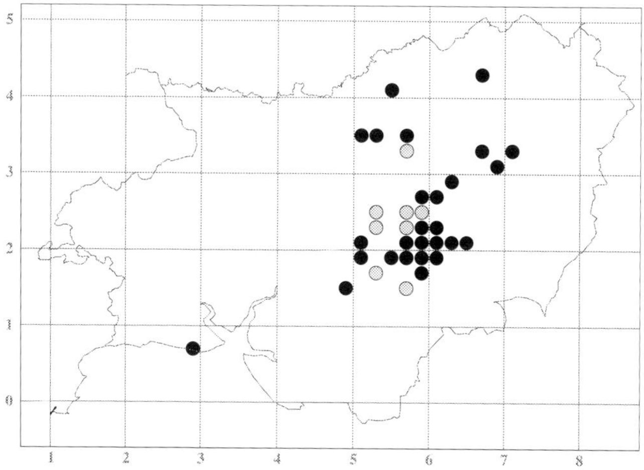

Fallow deer may have been native to Britain at one time, but became extinct after the last glaciation. The Normans reintroduced them in the 11th century and by the 14th century there were many hundreds of deer parks throughout Britain. It was from these that the landscaped deer park developed.

Mature broadleaved woodland is the preferred habitat of fallow deer, though they will also use conifers, but woods need not be large as they are only used for shelter. Fallow deer often forage on agricultural land and are unselective feeders with grasses, herbs and broadleaved browse forming the bulk of the diet. Man causes most mortality, though foxes may take newborns. Road casualties also account for many deaths, and males may occasionally die of injurics sustained during the autumn rut. Fawns are born in the spring and are suckled by the female until the following winter.

The introduction of fallow deer to Carmarthenshire dates from around 1600 at Golden Grove. Between 1565 and 1570 the first house was built on what was to become the estate, several hundred yards from the location of the current mansion. At some point in the next thirty years, a deer park was added, and fallow deer introduced. Initially a source of food, they soon became the privilege of the upper classes and very much a status symbol.

John Vaughan's notebook of 1755 (now in the Cawdor collection in the archive of Carmarthenshire County Council) records a total of 600 bucks, does and fawns in the park. Little note is made of their existence, apart from occasional references to their appearance on the table in various banquets.

Fenton (1917) records an incident during the Civil War when Oliver Cromwell visited Golden Grove. Cromwell was travelling from Glamorgan on his way to besiege Pembroke, a royalist stronghold at the time. Cromwell's intention was to capture Lord Carberry, a noted royalist, but Carberry was forewarned and, somewhat unchivalrously, fled into the mountains, leaving his wife to entertain Cromwell. There was no venison in the house, so Lady Carberry sent out for a buck to be shot. Cromwell is said to have shot the buck himself, and Fenton claims to have seen the tree where he rested his gun, though this detail appears to have been a local legend. Samuel (1868) quotes the Earl of Carberry saying that *... Cromwell sent from parks near London several staggs* [sic] *unto him to furnish his park at Golden Grove ...* in exchange for the one on which he had dined.

Until quite recently there was a captive herd below the mansion, with owners adding to the stock from the herds at Margam in Glamorgan. In about 1950, when the mansion was converted to a college, there were some 80 animals present in and around the park. Many escaped at this time, most subsequently being shot, but some were seen up to fifteen miles away (James, 1959).

Across the valley from Golden Grove, the Dinefwr Park herd also has its origins in the 1600s. Between 1590 and 1610, the Rice family began buying up land in what may have been an attempt to build up a consolidated estate, possibly to establish a deer park. By 1660, the house that still stands today had been established. Paintings from the period now hanging in Newton House show the presence of deer on the estate, some of which are identifiable as fallow deer. It appears that the deer park was established at about this time. In the late 1770s, when the development of natural rolling landscapes became fashionable, the estate was visited by Lancelot 'Capability' Brown, who was very impressed with Dinefwr. Brown suggested a few additions to the estate, including a design for deer park fencing. The importance of the deer to the estate economy is indicated by the addition of a deer abattoir at about 1790.

In the 1960s, the death in quick succession of Walter Rice and Charles Rhys, the seventh and eighth Barons of Dinefwr, and the consequent death duties, caused the sale of much of the estate. The deer park went into a period of decline, with animals regularly escaping and poachers coming into the park with dogs. Many animals would have dispersed into the surrounding countryside at this time. Since 1986, the park has been in the ownership of the National Trust, who are now engaged in its renovation (Dignam, undated).

During this century, fallow deer distribution has remained much as it is today, with expansion of the population held in check by shooting. This was particularly acute during the

Second World War when, with fresh meat in short supply, any deer wandering far from the Dinefwr estates was likely to end on the kitchen table.

Today, hunting still contains this attractive deer to a core area around the original introduction sites, although individuals have been seen as far as Brechfa Forest. There is also a small population of fallow deer on the dunes between Laugharne and Pendine. They are thought to have originated as escapes from a wildlife park in the area. The high security surrounding the Ministry of Defence establishment at Pendine affords them complete protection, and they can be expected to expand in this area in due course.

ROE DEER

Capreolus capreolus

IWRCH

Formerly extinct, now a recent recolonist from the north east.

Roe deer were common in Carmarthenshire up to the late mediaeval period but, following their extinction at that time, have begun recolonising woodland in the county in recent years.

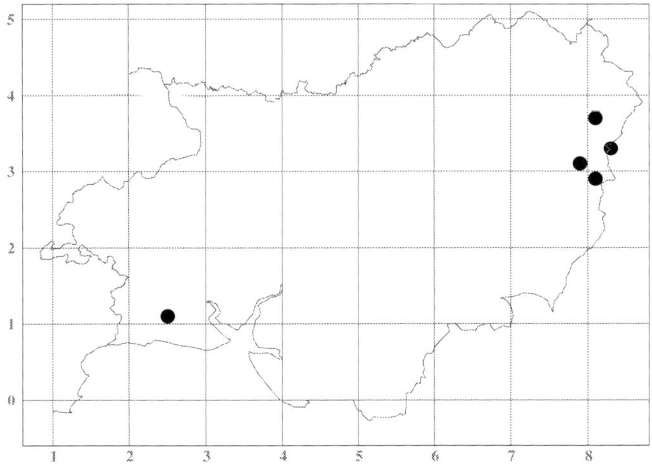

This small deer is usually encountered in coniferous or deciduous woodland, where it feeds on buds, shoots, herbs, bramble, ivy, conifers and ferns. Mating takes place from mid July until August when the rut occurs. Females give birth the following May or June to a single fawn, occasionally two and rarely triplets. Approximately 50 - 80% of yearlings are lost every year because of predation or emigration from the breeding area. Man is the main predator, killing more than 15,000 animals nationally per annum, although there is some predation of kids by foxes (Staines and Ratcliffe, 1991).

Roe deer would have been present in Carmarthenshire's woodlands from the earliest period. The first reference, as with so many species, comes in The Laws of Hywel Dda. The skin of the roe was declared to be the prerequisite of the cook or the king's steward, and was valued at one penny, the same as that for a sheep or a goat. This suggests that the roe deer was common at that time (Cadman, 1958; Matheson, 1932).

Evidence from later times is more scanty. An antler of a roebuck was unearthed during excavations at the mediaeval castle at Kidwelly and sent to Colin Matheson at the National Museum of Wales (Matheson, 1932). This castle was occupied from 1100 to 1300 and less intensively for two centuries afterwards. Presumably the antler was deposited at some point during this period, perhaps cast aside in the preparation of the meat for some feast. Roe deer were also mentioned in the forest law of the mediaeval Forest of Glyncothi, which stood roughly in the area occupied by Brechfa Forest today. The forest itself was created in about 1300, following the conquest of west Wales by Edward I. Roe deer are described as 'beasts of venison', the term at that time not referring to deer meat but being a technical term describing the game animals of the forest. There was even a peculiarly Welsh way of trussing, or 'hurdling' the animal once it had been killed (Linnard, 1984). Contemporary evidence of the presence of roe deer comes from a *cywydd* (see Chapter 2) written by the poet Lewis Glyn Cothi (1447-1487) to Rhys ap Dafydd. In it he describes roebucks 'by tens' in the idyllic forest of his youth. There is little doubt therefore that the species was widespread in Carmarthenshire woodlands up to this period (Rees, 1995).

The precise date of disappearance of roe deer in Carmarthenshire is uncertain, but is thought to be before 1600, for the roe was absent from neighbouring Pembrokeshire at that time (Matheson, 1932). What is certain is that the species has made a return since 1990. Two individuals were recorded near Pendine in 1990, thought to be escapes from a nearby wildlife park. The feral population, such as it was, does not seem to have persisted. At about the same time animals that have spread into Wales from introduced stock in England have begun to appear in the conifer forests of the north east of the county (Lucas, 1996). The species is now well established in that area, and whilst some tree damage has been reported, forestry workers do not consider it serious at present. As the population expands, however, attitudes are likely to change. Despite the difficulties this may cause, it is pleasing to see a long lost member of our mammal fauna return to take its place in the countryside of Carmarthenshire.

CHINESE MUNTJAC

Muntiacus reevesi

CARW MWNTJAC

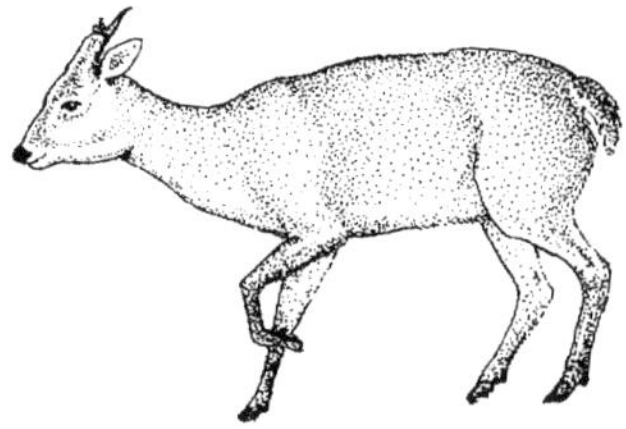

A recent coloniser in the north and west.

This diminutive deer is native to south east China and Taiwan, but is now widespread and increasing in southern Britain.

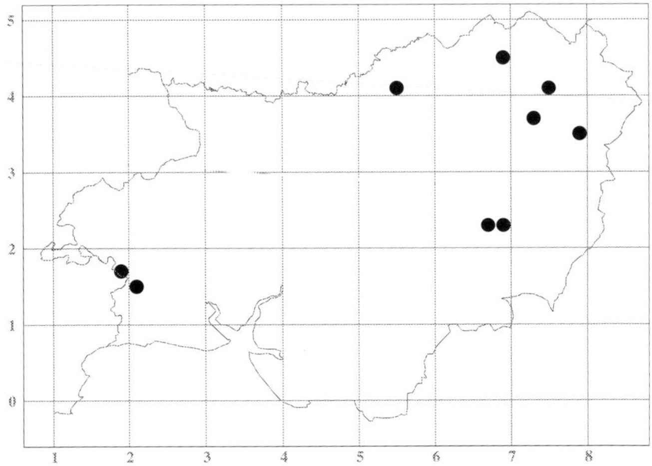

Chinese muntjac were originally introduced to Woburn Park in the early years of this century, and have spread to other parts of Britain from there. There have also been subsequent escapes from captive populations in Bedfordshire, Hertfordshire and Northamptonshire, as well as numerous undocumented introductions elsewhere. Chinese muntjac prefer dense habitat - coppice, spinneys and plantations, especially of conifers. They are solitary, with a range of about 14 ha, and can breed throughout the year. Females are therefore almost continuously pregnant. Chinese muntjac are browsers, feeding on brambles, ivy, ferns, fungi, leaves, nuts, fruit and grasses. Road casualties are a major cause of mortality, although shooting and foxes also take a toll (Chapman, 1991).

Chinese muntjac are comparatively recent arrivals in Carmarthenshire, having almost simultaneously appeared at opposite ends of the county. The species has been recorded across the border in Pembrokeshire, where a population is believed to have become established, partially as a result of three animals said to have escaped from a now closed wildlife park at Pendine (Chapman *et al.*, 1994). This has lead to a cluster of records in the south west of the county. Meanwhile, during the 1980s or early 1990s, it is presumed that there was a release in the Llwyndrissi area, leading to records in the Llanybydder area. However, there have also been occasional reports from Crychan Forest and the Rhandirmwyn area. This suggests that at least some of the records for Carmarthenshire originate from genuine colonisation from adjacent counties rather than animals being moved in by human agency. Chinese muntjac have been recorded as road casualties, although at least one was found part eaten by a fox. In 1994 one was recorded as far south as Golden Grove, and in 1995 another was seen near Llandeilo (Lucas, 1996).

Chinese muntjac are now widespread, but local, moving mainly along the river valleys, and can be expected to colonise much of the county in due course.

CHAPTER 5:
EXTINCT MAMMALS IN CARMARTHENSHIRE

PREHISTORIC MAMMALS

What is known of prehistoric mammals in the county originates from one site. Coygan Cave stood a few miles to the west of Laugharne, overlooking Carmarthen Bay, in an outcrop of Carboniferous Limestone. It was a large cave, consisting of three main passages converging on a central chamber. It was formed by the action of rain water, though waves could have expanded the cave at a time when the sea level was higher.

The cave was excavated several times between 1866 and 1933. From the abundant coprolites (fossil dung) found in the site, it appears to have been a hyaena den. Scattered throughout the cave were bones of the animals from the surrounding countryside, on which the hyaenas preyed. At that time, some 40,000 years ago, the sea level was much lower than it is today. Carmarthen Bay would have been a broad grassy plain, with scattered coniferous woodland on the hills to the north. As a result, Coygan Cave yielded almost the complete suite of what has been called the 'Pleistocene Megafauna' (Grimes and Cowley, 1935; Clegg, 1969).

Among the species recorded were wild horse *Equus caballus*, cave hyaena *Hyaena crocuta* var. *spelaea*, woolly rhinoceros *Rhinoceros tichorhinus*, mammoth *Elephas primigenius*, giant ox *Bos primigenius*, Irish elk *Cervus giganteus*, cave bear *Ursus spelaeus*, reindeer *Rangifer tarandus* and neanderthal man *Homo neanderthalensis.*

Readers may be puzzled by the persistent use of the past tense in referring to this site. Inexcusably, the cave was completely quarried away in the late 1970s, an act of unbelievable archaeological vandalism. Given the current interest in the wildlife and archaeology elsewhere on the limestone ridge of Carmarthenshire, it is to be fervently hoped that similar action is not repeated.

SHIP RAT

Rattus rattus

LLYGODEN DDU

The globally common ship rat now ranks as one of Britain's most endangered mammals (Morris, 1993). Formerly widespread in Britain, since 1984 the only mainland records have been from London, though short lived colonies do occur at ports as a result of accidental introductions.

The ship rat is thought to have arrived in Britain in Roman times, and was widespread in mediaeval Britain. The first reference to the ship rat in Wales is contained in *The Journey through Wales* of Gerald of Wales (Thorpe, 1978), in which he mentions a man being devoured by rats. The ship rat has been widely, if erroneously, blamed for the spreading of the plagues of the 13th and 17th centuries (Twigg, 1984). During the Black Death of the 14th century, twenty-five million people were said to be killed in Europe, a quarter of the population. In 1349, Carmarthen was mentioned as a location where the plague was particularly bad (Matheson, 1931, 1955).

During the late 18th and early 19th century, the ship rat disappeared from all of inland Wales, and in 1796 Pennant reported that the brown rat had ... *destroyed the Black* (i.e. ship*) Rat in most places.* (Condry, 1981). In 1945 it was still present at Newport, Cardiff, Swansea, Milford Haven, and Holyhead, but by 1951 only Cardiff and Swansea had surviving populations. A decade later it was thought to be confined to Cardiff (Hill, 1961). The replacement of the ship rat by the brown rat has been attributed to the brown's larger size and its ability to withstand greater climatic extremes.

The recent distribution of ship rats in Carmarthenshire is largely a matter of speculation. Given the species' confinement to busy ports, and the survival of Llanelli as a shipping centre until the early half of this century, it is possible that the species may have lingered intermittently in that area alongside more well known centres to the east. If true, then the ship rat has the dubious distinction of being the only mammal species to have become extinct in Carmarthenshire in the 20th century. Following the decline of the port of Llanelli, the ship rat is now unlikely to recolonise the county. Environmental health officers report no occurrences in the last few decades.

BEAVER

Castor fiber

AFANC

Beavers would have been numerous in the wooded valleys of Carmarthenshire before hunting took its toll. One of the few references to the beaver in the county comes in the Laws of Hywel Dda (Condry, 1981) which describes the value of a pelt as being 120 pence each, compared to the pine marten at 24 pence and the otter a mere 8 pence. Condry, in fact, doubts whether the furs referred to are of Welsh origin, but all the other species mentioned were indigenous to Carmarthenshire at that time. It is just possible that the beaver could have clung on into the 10th century, the high value of the pelt reflecting its rarity even then.

The beaver has a special place amongst extinct mammals in the county because of an extensive account of the beaver population on the Afon Teifi, near Cenarth, by Gerald of Wales. He claimed that the Teifi was the last haunt of the beaver in Britain, and that it was still present in 1188. In fact, beavers were almost certainly present on other rivers in Britain, indeed several place names in north Wales include the word 'Afanc'. Condry again regarded the story as doubtful, but the extensive description of the beaver given by Gerald, which continues for several hundred words, repeated in *The Journey Through Wales* and *The Description of Wales*, appears to have been written in part from his own observations. In any event, the species must have been on the brink of extinction, for it seems to have disappeared shortly after Gerald wrote his account.

WOLF
Canis lupus
BLAIDD

No other animal has such a clouded history in Carmarthenshire than that of the wolf. Until humans began to compete with the wolf, it would have been the most widely distributed of all carnivores. Yet, although it was the progenitor of that most ubiquitous companions of man, the dog, wolves have been relentlessly hunted for hundreds of years.

The wolf was recorded from cave sediments excavated from Coygan Cave near Laugharne, along with other elements of the Pleistocene fauna such as Irish elk and bear. It has also been recorded from the caves on nearby Gower from the same period, suggesting that the species was indeed numerous throughout Wales before historic times (Matheson, 1932).

Little is known of the wolf in the early historical period. Lloyd - James (1909) reports that wolves harried the horses used in the lead mines in the Taf valley in Roman times, which had to be sheltered from them nightly. He also noted a local name 'Blaiddbwll' - the Wolf's Pit. Wolves appear from the mists of time again towards the end of the 10th century. The Laws of Hywel Dda gave no legal worth to the wolf and it was 'free to all to slay them'. Both Harting (1880) and Matheson (1932) record a story of how the Welsh were compelled to pay taxes to the Saxon kings in the form of wolves heads, which at that time plagued the countryside. About the year 953, the Saxon king Athelstan entered Wales with an army, using as a pretext the murder of one of his vassals in an area occupied by the Prince of Dinefwr. He then imposed a levy of £20 in gold, £300 in silver and 200 head of cattle. This levy was paid in part by the Prince of Dinefwr, in whose land the crime had occurred. For reasons that are not altogether clear, the levy was commuted to a tax paid in wolf heads and pelts. The tax is said to have existed nominally (for apparently the Welsh often refused to pay it) for thirty-two years, from the time of Athelstan to Edgar. In fact, the historical truth of this story is in some doubt as it only appears at a much later date in the 12th and 13th centuries (H. James, pers. comm.). However, it should be noted that the tale would have made little sense to the inhabitants unless wolves were present at the time.

Persecution seems to have had limited long term effect on their numbers; in 1166, according to the ancient chronicle known as the 'Annales Cambriae', a rabid wolf ran amok in Carmarthen, biting eighteen people, almost all of whom died.

The precise date for the extinction of the wolf is uncertain and shrouded in legend. Congreve (1927) gives a date of 1576 for the whole of Wales. The wolf could have clung on in north Carmarthenshire, which would have been even more remote than it is today, until about this time. Pocock (1927) describes three alleged wolf pads found in Island House in Laugharne said to have come from the last wolf to be killed in the area. It apparently met its fate near Marros, and he regarded the pads to be genuine. Turner-Evans (1962) noted a place near Blaen-waun called Pantyblaidd, where the last Welsh wolf was supposedly killed during the reign of Henry VIII. A local monument is reputed to mark the spot, although the structure, in fact, dates from the Stone Age.

WILDCAT

Felis sylvestris

CATH WYLLT

Records of 'wildcats' are complicated by the confusion by early writers between the genuine wildcat and feral domestic cats.

Notwithstanding the claim by Fenton (1806) that the last wildcat in Wales was killed at Llethr Mawr, Taliaris, in about 1650, wildcats are frequently mentioned in the vestry records of various parishes. Brunker (*c.* 1950) quotes the vestry book in Llanegwad from 1732, which notes the reward for killing a wildcat as one shilling. Matheson (1932) claimed that the wildcat survived in small numbers in Pembroke and Glamorgan until the early years of the 19th century, and the species presumably also remained in Carmarthenshire at that time. As to the precise date of extinction, Langley and Yalden (1977) assumed that it disappeared from Carmarthen, along with other counties in south Wales by 1850 at the latest, ... *retreating to the wilder parts of the remaining counties.* However, northern Carmarthenshire would have been as wild as any other part of Wales at that time - witness the survival of the red kite in the upper Tywi valley. It is just possible that the wildcat could have lingered here. Bell (1874) notes the wildcat as being restricted to the woody mountains of Wales, just the kind of habitat present in the north of Carmarthenshire. In any event, it did not last long, and probably became extinct between 1850 and 1860. There is however an intriguing note of two animals, showing some characteristics of wildcats, being shot near Hebron in 1931.

It is thought provoking to compare the fate of the wildcat with two other large carnivores, the polecat and the pine marten. In both these cases, the First World War arrived just in time to release them from the relentless persecution of Edwardian Britain. Had the wildcat clung on for another 50 years, we might today be talking about the recovery of this species and planning its conservation, rather than regarding it, like the wolf and the bear, as belonging to a bygone age.

BROWN BEAR

Ursus arctos

ARTH

The brown bear would have been widespread in the forests of Carmarthenshire were it not for the persecution by man. It appears to have been one of the first species to have become extinct within historic times, probably disappearing before 1066 (Linnard, 1982). There is little information on its distribution in Carmarthenshire, although some (e.g. Harting, 1880) have claimed that the presence of the word 'Arth' in many place names may indicate its former distribution. This is fanciful. Sadly, the word also means an enclosed place and is one of the plethora of Welsh words for 'hill', a rather more plausible explanation of its use in names such as Llanfihangel ar Arth.

RED DEER

Cervus elaphus

CARW COCH

Though red deer are nowadays associated with the wide expanses of the treeless uplands of northern and western Britain, the species was originally a woodland animal, found throughout Wales. Gerald, in *The Journey Through Wales,* mentions red deer in terms suggesting that they were very abundant. Matheson (1932) reports an antler from material dating from the 13th century, unearthed from Kidwelly Castle. A report in the *Transactions of the Carmarthen Antiquarian Society* (Anon., 1923) reports a red deer antler 27 inches long of unknown date being found near Pensarn brickyard, some twelve feet below the surface. The abundance of the word *Carw* in Carmarthenshire place names (see Appendix 3) also suggests the species was very common at least up to mediaeval times.

Red deer also represented a valuable source of protein to the hungry Welsh, and were hunted to extinction. No precise date for its demise is known, though Matheson (1932) quotes the close of the 18th century as the extinction date for the whole of Wales.

Eight red deer were apparently introduced into Dinefwr Park in 1975, but all were shot shortly afterwards.

WILD BOAR

Sus scrofa

TWRCH

The wild boar is an inhabitant of deciduous woodland, and would have been widespread in the county before the clearance of the wildwood.

Wild boar are mentioned in the Laws of Hywel Dda (Jenkins, 1990), where the season for boar hunting is given as between 9 November and 1 December. Forest laws appertaining to the Forest of Glyncothi (Rees, 1995) suggest that the species was a common subject of the chase in that area until the late Middle Ages.

The precise date of extinction is uncertain, as the species was also domesticated out of existence (modern varieties of pig are a comparatively recent phenomenon - domestic swine in the middle ages looked little different from their wild counterparts). However, the late 16th century is the most likely date, when large scale deforestation began.

A photograph of a sculpture of a boar was published in the *Transactions of the Carmarthen Antiquarian Society* (Vol. 10, p 62) taken at Edwinsford Hall. A skull, reputed to be a hunting trophy, was also found in Island House, Laugharne (Pocock, 1927).

APPENDIX 1: POTENTIAL FUTURE ADDITIONS TO THE CARMARTHENSHIRE LIST

It is occasionally entertaining to speculate on species yet to be discovered in Carmarthenshire. The species below are thought to be the most likely future additions to the county list.

Bechstein's bat

Myotis bechsteinii

Ystlum Bechstein

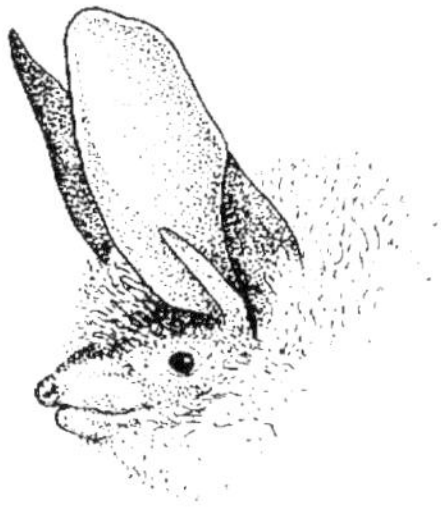

A species similar to Natterer's's bat, mainly recorded in southern England the Welsh marches. Its habit of normally roosting in trees throughout the year means that the species could occur in Carmarthenshire but remain undetected.

The extensive wooded valleys of the county may provide suitable habitat.

Serotine

Eptesicus serotinus

Ystlum Adain-Lydan

The serotine is a large species of bat, found almost exclusively in houses. The main area of distribution in Britain is in the south east, from Suffolk to Dorset, although there are isolated records from Wales.

Mathew (1991) notes an unconfirmed record from the county. T.McOwat has made field identifications of the species near Llandeilo and Gwernogle.

Leisler's bat

Nyctalus leisleri

Ystlum Leisler

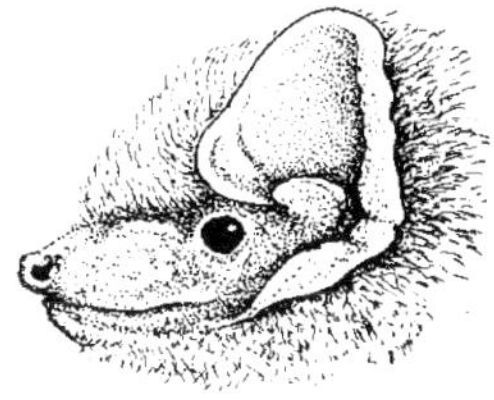

Leisler's bat is a large species, slightly smaller than the very similar noctule. It is widely but thinly distributed in England and the Welsh borders. However, in Ireland where the noctule is absent, it is the third most common species.

A bat, possibly of this species, was seen near Abergwili in 1996.

Barbastelle

Barbastella barbastellus

Ystlum Du

This small black bat is rare throughout Britain, with scattered records south of a line from the Tees to the Mersey, including records from east Wales and Pembrokeshire (Arnold, 1993). The species is known to breed in Norfolk.

The species is known to be associated with wooded valleys, a habitat Carmarthenshire has in abundance.

Nathusius's pipistrelle

Pipistrellus nathusii

Ystlum Lleiaf Nathusius

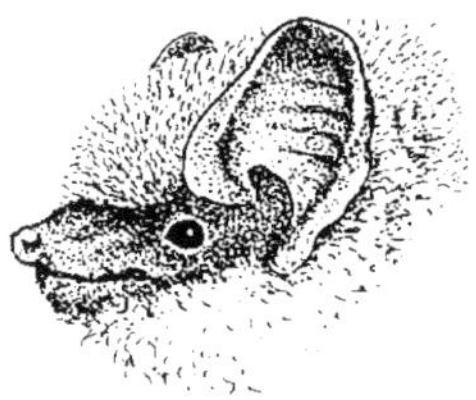

This European bat was proved to be breeding in Britain for the first time in 1997 at two locations, in Northern Ireland and Lincolnshire. Prior to that year, there were records for some 50 individuals of this species in Britain (Hutson, 1997). An individual was identified by Dr Bob Stebbings during a training course at Stackpole, Pembrokeshire, in September 1994. This was the first record for Wales and the thirteenth for Britain (A. Poole, pers. comm.).

Bats that, in retrospect, showed some characteristics of Nathusius's pipistrelle have been seen in the Mynydd Du area. With the greater awareness that Carmarthenshire batworkers now have for this species, it may be only a matter of time before it is added to the county list.

Risso's dolphin

Grampus griseus

Dolffin Risso

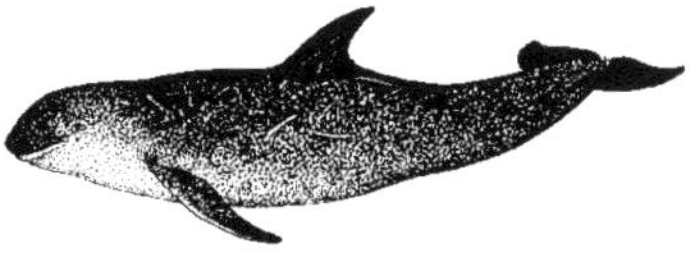

A large dolphin which occurs primarily on the Atlantic coasts of the British Isle and is considered regular in the Irish sea (Evans), 1991). The species has been regulary recorded in neighbouring Pembrokeshire.

Sika deer
Cervus nippon
Carw Sica

This an introduced species, native to Japan. In Britain, it occurs as the result of escapes or deliberate introductions in southern England, the north of England, the southern uplands of Scotland and the Highlands.

Deer of this species are reputed to be present on Laugharne-Pendine Burrows, although all photographs thus far have proved to be fallow deer. The nearest population is near Cardigan, Ceredigion.

APPENDIX 2: ESCAPES AND EXOTICS

Racoon

Procyon lotor

Racwn

Racoons are native to North and Central America, and are familiar animals of woodland and parkland. The fore-paws are often used as 'hands' for holding food, giving it an attractive nature which no doubt adds to its appeal as a zoo animal and unusual pet.

One individual was seen near the Usk reservoir in about 1992, by forestry workers. Interestingly, this is not far from another known escape site across the border near Brecon (Hills, 1991).

Coypu

Myocaster coypus

Coipw

The coypu is native to South America, where it inhabits swamps and rivers but was imported to Britain, from about 1929, for fur farming. A population was established in East Anglia, but following an eradication campaign, is now believed to be extinct.

An individual of this species recorded on the Afon Teifi is noted by Holiday (1961).

Finally, there is a note in the minutes of the Llangennech Parish Council for sometime in the mid 1930s though it has not been possible to research this reference further. The true identity of the animal, or whether it was a hoax, may never be known, and the author offers the passage in its entirety:

> *On a lighter note, on 1st November, one of the local newspapers, recorded that: 'Mr William John Evans of Glanmwrwg Fach Farm, Llangennech, alerted yesterday by the barking of dogs; discovered a freak animal, in the farmyard. It had the head of a badger, the feet and body of an otter and the fur of a rat, its length from head to tail being 42 inches. It was a foot high and weighed 12lbs. Mr Bert Carr, a well known showman who is present at Llangennech Fair after seeing the beast said that he was of the opinion that it was a cross between a badger and an otter. It was the first of its kind he had ever seen'.*

APPENDIX 3: MAMMALS IN CARMARTHENSHIRE PLACE NAMES

Welsh is a language of great antiquity. Prior to the Roman Conquest, a Celtic language known as Brythonic or Brittonic was spoken throughout what is now southern Scotland, England and Wales. It is from this that Welsh and the closely related Cornish languages are descended.

One feature of Welsh is its stability compared to other languages. For example, Chaucer's poetry of the 14th and 15th century has to virtually be translated into modern English, but his Welsh contemporary, Dafydd ap Gwilym, is still comprehensible to the modern Welsh speaker. Indeed, a poet writing today would use similar Welsh if writing the same kind of poem.

The same is true of place names, for whilst names in England such as London, Birmingham or Manchester have no clear meaning, most Welsh place names are intelligible today. Working in Carmarthenshire, the author was struck by the number of occasions on which mammals appeared in place names. Fortunately, the Cambrian Archaeological Trust have a database of place names in the county drawn from the first edition of the six inch Ordnance Survey, tithe maps, and the 1948 1:25000 Ordnance Survey maps. This database was interrogated for all mammal species names, together with others such as the Welsh equivalents of 'hind' and 'doe'. Non-Welsh speakers should be aware that, in Welsh, beginnings of words can change as well as endings; the 'mutations' with which learners have so much difficulty. The reasons for this need not trouble us here, but the reader needs to appreciate that, for instance, 'Carw' and 'Garw' both mean deer, and are not spelling errors. A guide to pronunciation can be found in any text for Welsh learners or dictionary.

The results are as follows, with English translations. Where a site has an anglicised version of the name in current usage, this is given in brackets after the Welsh name.

Ysgyfarnog (hare)

Cwmysgyfarnog	SN568232	Hare Valley
Allt Cwmysgyfarnog	SN566231	Hillside of the Hare Valley
Nant Ysgyfarnog	SN694477	Hare Stream

Llygod (rat, mouse or vole)

Pwll Llygod	SN446068	Rat Pool

Whilst llygod is used for rat or mouse in common parlance, could the connection with a pool indicate water vole?

Cadno (fox, in the South Wales dialect)

Nant Cadno	SN636099	Fox Stream
Blaen Nant Cadno	SN646097	Source of the Fox Stream
Nant y Cadno	SN639096	The Stream of the Fox
Bryncadno	SN604263	Fox Hill
Cwm Cadno	SN598129	Fox Valley
Cwm y Cadno	SN815310	Valley of the Fox
Pont Allt y Cadno	SN453137	Bridge of the Hillside of the Fox
Allt y Cadno	SN456136	Hillside of the Fox
Allt Cwm Cadno	SN631405	Hillside of Fox Valley
Cwrt y Cadno	SN692441	Court of the Fox
Cadno	SN236088	Fox

Bele (pine marten)

Aber Bele	SN440350	Confluence of the Marten (river)
Felinbele	SN440353	Marten (river) Mill
Bryn Bele	SN442351	Marten Hill
Nant Bele	SN443344	Marten Stream
Blaen Bele	SN448336	Source of the Marten Stream
Clyn Bele	SN343331	Marten Valley
Afon Bele	SN356332	Marten River
Wern y Pele	SN290362	Marten Swamp

Near Blaen Bele there is also Blaen Bele Uchaf and Blaen Bele Isaf - Upper and Lower Blaen Bele.

Ffwlbart (polecat)

Bryn Ffwlbart	SN376290	Polecat Hill

Dwrgi (otter)

Cwm Dwrgi	SN313226	Otter Valley
Blaen Dwrgi	SN358244	Source of the River of the Otter
Llety'r Dwrgi	SN421182	Otter Inn
Lletty'r Dwrgi	SN465273	Otter Inn
Penlandwrgi	SN465274	Parish Hill of the Otter
Ffynnon Dwrgi	SN465274	Spring of the Otter

Dwrgi literally means 'water dog'.

Carw (deer; this word is usually used in relation to red deer)

Llwyn Garw	SN622150	Deer Bush
Cae Garw	SN638159	Deer Field
Llethr Garw	SN609190	Deer Slope
Cae Garw	SN657223	Deer Field
Tyddyn Cwm Garw	SN721143	Deer Valley Farm
Cwmgarw	SN721143	Deer Valley
Blaen-Cwm-Garw	SN721153	Head of the Deer valley
Nant Garw	SN724155	Deer Stream
Pen-Cae-Garw	SN703195	Deer Field Hill
Pen-Cae Garw	SN736199	Deer Field Hill
Pen-y-Garw	SN736199	Hill of the Deer
Garwnant	SN792236	Deer Stream
Allt-y Carw	SN775281	Hillside of the Deer
Bwlch-Garw	SN553137	Deer Pass
Llwyn-Carw	SN582183	Deer Bush
Castell Garw	SN153269	Deer Castle
Cilcarw	SN483106	Deer Corner
Cae-Garw	SN498122	Deer Field
Bwlch y Garw	SN454170	Pass of the Deer
Pen-Bryn Garw	SN477243	Deer Hilltop
Bryn-Garw	SN478242	Deer Hill
Glan-Carw	SN499247	Deer Riverside
Waun Garw	SN209155	Deer Meadow
Parcgarw	SN368152	Deer Park
Carwe (Carway)	SN465067	Deer Place
Cae-Garw	SN466093	Deer Field
Pant y Garw	SN296378	Deer Valley
Glyn Garw	SN205274	Deer Valley
Pen-Clyn-Garw	SN211275	Deer Valley Hill

The aggregation of names with Carw elements in the Gwendraeth Valley area, in the former Forest of Glynne, is particularly striking.

Hydd (stag)

Pentre'r hydd	SN384210	Village of the Stag

Iwrch (roe deer)

Craig'r Iwrch	SN770162	Roe Deer Crag
Nant-Iwrch	SN706394	Roe Deer Stream
Cwmiwrch	SN648358	Roe Deer Valley

Ewig (roe deer hind)

Pant yr Ewig	SN506250	Valley of the Roe Deer
Waun yr Ewig	SN523240	Meadow of the Roe Deer
Ffos yr Ewig	SN571270	Ditch of the Roe Deer
Clyn yr Ewig	SN541373	Valley of the Roe Deer
Waun yr Ewig	SN540374	Meadow of the Roe Deer

Cath (cat)

Waun-Clyn-Cath	SN634136	Cat Valley Meadow
Cwmnantygath	SN684130	Valley of the Stream of the Cat
Carregcath	SN699246	Cat Rock

Names with the element 'cath' need interpreting with care, for the cat referred to may not have been a 'wild'cat.

Twrch (boar)

Twrch Fechan	SN783203	Little Boar (River)
Cefn Twrch	SN807213	Boar Ridge
Afon Twrch	SN784193	Boar River
Twrch Fechan	SN784193	Little Boar (River)
Cwm Twrch	SN497220	Boar Valley
Ffynnon-Twrch	SN495220	Boar Spring
Carn Twrch	SN809462	Boar Cairn
Ffrydiau Twrch	SN770162	Boar Hill Pasture
Cwm Twrch	SN757131	Boar Valley
Pwll Twrch	Near Kidwelly	Boar Pool
Pant Twrch	SN493137	Boar Valley
Blaen-Twrch	SN681495	Head of the Boar Valley
Craig Twrch	SN661492	Boar Crag
Glan Twrch	SN646411	Boar (river) side
Afon Twrch	SN647447	Boar River
Cwm Twrch	SN652474	Boar Valley
Craig Twrch	SN653481	Boar Crag

Most of these names owe their origins to two rivers named Twrch. No grid reference was available for Pwll Twrch.

Blaidd (wolf)

Nant-y-Blaidd	SN372299	Stream of the Wolf
Cil-Blaidd	SN541459	Wolf Corner

Cae-Blaidd	SN524336	Wolf Field
Cae'r Blaidd	SN518334	Field of the Wolf
Cefn-Blaidd	SN647334	Wolf Ridge
Tre-Bleiddiau	SN226241	Wolves' Village
Nant y Fleiddiast	Near Rhandirmwyn	Stream of the Wolf Bitch

No place names were located for the following mammal names; Twrch Ddaear/Gwadden (mole), Ystlum (bat), Gwiwer (squirrel), Pathew/Pathor (common dormouse), Carlwm (stoat), Gwenci/Bronwen (weasel), Mochyn Daear (badger), Danys (fallow deer) and Afanc/ Llostlydan (beaver). A number of places were located with the name Draenog (hedgehog), but as this is also means 'thorny' it is unlikely that these have a mammal connection.

APPENDIX 4: NATIONAL GRID REFERENCES OF CARMARTHENSHIRE SITES MENTIONED IN THE TEXT

Abergwili	SN 437211
Abermarlais Park	SN 694297
Abernant	SN 339232
Afon Bran	SN 773343 - 815450
Afon Cothi	SN 515229
Afon Tywi	SN 355095 - 794485
Allt Rhyd y Groes	SN 764480
Alltwalis	SN 446317
Aman	SN 623106 - 752179
Aman Valleys	SN 61
Ammanford	SN 630124
Amroth	SN 164071
'Ashpits Pond'	SN 465010
Bannau Sir Gaer	SN 810214
Beddau'r Derwyddion	SN 605165
Bevan's Pond	SN 414045
Bishop's Pond	SN 444210
Blaen-waun	SN 237273
Bone Cave	SN 21-08-
Brechfa	SN 525304
Brechfa Forest	SN 43, 53
Burry Inlet	SS 49
Burry Port	SN 445008
Bynea	SN 456994
Caeo Forest	SN 743410
Cae'rbryn	SN 598138
Capel Dewi	SN 476203
Carmarthen	SN 410200
Carmarthen Bay	SN 00
Carreg Cennen	SN 667194
Castle Woods	SN 613217
Cefn Farm	SN 812387
Cefn Sidan	SN 00
Cenarth	SN 267415
Cilycwm	SN 753400
Cothi	SN 501201 - 703487
Cothi Valley	SN 52, 53
Coygan Cave	SN 286093
Croesyceiliog	SN 407165
Cross Hands	SN 561127
Crwbin	SN 473133

Crychan Forest	SN	83
Cwm Morgan	SN	294349
Cwrt Henri	SN	556227
Cwrtycadno	SN	692442
Cych	SN	235414 - 290326
Cydweli	SN	409067
Cynghordy	SN	817398
Cynin	SN	283160 - 273300
Cynwyl Elfed	SN	374275
Cywyn	SN	307129 - 316298
Dafen Brook	SN	526006
Derwydd	SN	625180
Derwydd Mansion	SN	613178
Dinefwr	SN	62, 73, 74, etc.
Dinefwr Oxbows	SN	616224
Dinefwr Park	SN	615225
Drysgol	SN	685155
Dryslwyn	SN	554205
Duad	SN	384265 - 382337
Dynefor	SN	615225
Felinfoel	SN	524026
Felingwm	SN	508247
Ferryside	SN	366104
Fforest	SN	583046
Ffrwd Fen	SN	418026
Foel Fawr	SN	735185
Four Roads	SN	447094
Gelli Aur	SN	597198
Ginst Point	SN	332079
Glanrannell	SN	650372
Glasfynydd	SN	82
Glyn Abbey	SN	448079
Glynea Pond	SS	554989
Golden Grove	SN	597198
Gorlech	SN	586335 - 643386
Gwendraeth Estuary	SN	00
Gwendraeth Fach	SN	397066 - 567174
Gwendraeth Fawr	SN	409056 - 567137
Gwenffrwd/Dinas	SN	780465
Gwili (Llanelli)	SN	583032 - 572115
Gwili (North)	SN	432205 - 489304
Halfway	SN	828329
Harford	SN	637433
Henllys near Cilycwm	SN	756366
Kidwelly	SN	409067
Kidwelly Castle	SN	409071
Kymer's Canal Pond	SN	565058

Lampeter	SN	577482
Laugharne	SN	302106
Laugharne Burrows	SN	20
Llanarthne	SN	533202
Llanboidy	SN	217232
Llanddowror	SN	255145
Llandeilo	SN	630225
Llandissilio	SN	123215
Llandovery	SN	765345
Llandybie	SN	617154
Llanegwad	SN	518215
Llanelli	SN	510005
Llanelli 'Levels'	SS	52-98-, 53-98-, 54-98-
Llanfihangel ar Arth	SN	454393
Llangadog	SN	706284
Llangennech	SN	560015
Llangunnor	SN	430202
Llangynog	SN	339163
Llannon	SN	539085
Llansadwrn	SN	696315
Llansawel	SN	720363
Llansteffan	SN	352108
Llanybydder	SN	524440
Lliedi	SN	51-04-, 51-03-
Llwyn-teg	SN	552081
Llwyndrissi	SN	563415
Llyn Brianne	SN	795490
Llyn Llech Owain	SN	568152
Llyn Pencarreg	SN	537457
Llyn y Fan Fach	SN	803218
Loughor	SN	565005 - 668179
Machynys Point	SN	515973
Machynys Ponds	SN	512980
Manordeilo	SN	670260
Marlais	SN	696288 - 675347
Marros	SN	207089
Meidrim	SN	293208
Melinddwr	SN	623362 - 573380
Moat Farm	SN	433054
Morlais	SN	573016 - 512093
Mynydd Du	SN	71
Mynydd Llanllwni	SN	507380
Mynydd Mallaen	SN	74
Mynydd y Betws	SN	67-10-
Mynydd y Garreg	SN	442093
Mynydd Ystyfflau Carn	SN	470265
Nant Bargod	SN	354402 - 373340

Nantgaredig	SN 494213
Newcastle Emlyn	SN 306405
Newton House	SN 615225
'Oaklands', Cwmffrwd	SN 424166
Pembrey Forest	SN 00
Pembrey Motor Racing Centre	SN 40-03-
Pencader	SN 445360
Penclacwydd	SN 542983
Pendine	SN 233080
Penrhyngwyn Pond	SN 518975
Pensarn	SN 41-19-
Penybanc	SN 616241
Penygroes	SN 587133
Pibwrlwyd Farm Institute	SN 412183
Pinged	SN 424038
Pontardulais	SN 595035
Pumpsaint	SN 658404
Pwll	SN 476012
RAF Pembrey	SN 36-05-
Rhandirmwyn	SN 786436
Rhos Pwll y Gawnen	SN 294300
Rhos Wen	SN 480330
Rhydargaeau	SN 438264
Rhydcymerau	SN 578387
River Cennen	SN 633217 - 690196
Salmon Scar	SN 35-07-
Sandy Water Park	SN 494005
Sawdde	SN 694282 - 804220
St. Ishmaels	SN 363081
Stony Drain	SN 412056 - 419041
Stradey Woods	SN 48-01-
Sugar Loaf	SN 835427
Swan Pool Drain	SN 41-01-
Swiss Valley	SN 51-03-
Taf	SN 280153 - 215303
Tair Carn Isaf	SN 684168
Taliaris	SN 637287
Talley	SN 636326
Talley Lakes	SN 633334
Talyllychau	SN 636326
Tavernspite	SN 184127
Teifi	SN 246415 - 593481
'Three Rivers'	SN 00
Tir Philip	SN 47-18-
Trap	SN 653190
Trawscoed	SN 565376
Trelech	SN 283305

Trevaughan	SN 401214
Tumble	SN 540120
Twrch	SN 772085 - 816214
Tywi Forest	SN 75, 85
Tywi Valley	SN 42, 52, 62
Tywyn Point	SN 357064
Usk Reservoir	SN 820285
Warren Wood	SN 605192
Wharley Point	SN 340092
Whitland	SN 199165
Whitland Abbey	SN 208182
Witchett Pool	SN 275075
Ystrad House	SN 745329
Ystradffin	SN 787467

APPENDIX 5: SCIENTIFIC NAMES OF SPECIES OTHER THAN MAMMALS MENTIONED IN THE TEXT

PLANTS

alder	*Alnus glutinosa*
ash	*Fraxinus excelsior*
bilberry	*Vaccinium myrtillus*
bramble	*Rubus fruticosus*
dwarf willow	*Salix herbacea*
hawthorn	*Crategus monogyna*
hazel	*Corylus avellana*
heather	*Calluna vulgaris*
ivy	*Hedera helix*
mossy saxifrage	*Saxifraga hypnoides*
oak	*Quercus* spp.
roseroot	*Rhodiola rosea*
whorled caraway	*Carum verticillatum*
willow	*Salix* spp.

INVERTEBRATES

cockchafer	*Melolontha melolontha*
crayfish	*Austropotamobius pallipes*
dor beetle	*Geotrupes stercorarius*
earthworm	*Lumbricus terrestris*
freshwater shrimp	*Gammarus* spp.
house cricket	*Acheta domesticus*
marsh fritillary	*Eurodryas aurinia*
water louse	*Asellus aquaticus*
woodlouse	*Armadillidium vulgaris*

FISH

great white shark	*Carcharodon carcharias*
mackerel	*Scomber scombrus*
pike	*Esox lucius*

AMPHIBIANS

frog	*Rana temporaria*

BIRDS

barn owl	*Tyto alba*
bittern	*Botaurus stellaris*
buzzard	*Buteo buteo*
caspian plover	*Charadrius asiaticus*
dipper	*Cinclus cinclus*
grey heron	*Ardea cinerea*
long-eared owl	*Asio otus*
marsh tit	*Parus palustris*
red kite	*Milvus milvus*
short-eared owl	*Asio flammeus*
slender-billed curlew	*Numenius tenuirostris*
spoonbill	*Platelea leucorodia*
tawny owl	*Strix aluco*
willow tit	*Parus montanus*

REFERENCES

Anderson, S.S. (1991). The grey seal. In Corbet G.B. and Harris S. (ed) *The Handbook of British Mammals* 3rd ed. Blackwell Scientific Publications.

Anon (1883). Chased by Otters. *Llanelli Guardian* July 26 1883

Anon (1923). Red Deer Antler. *Transactions of the Carmarthenshire Antiquarian Society* 16: 65

Anon (1964). Mammal notes. *Nature in Wales* 9: 90-93

Anon (1970). Mammal notes. *Nature in Wales* 12: 110-111

Anon (1971). Killer Whale washed up at power plant. *Llanelli Star* 4.9.71

Anon (1997). *Action for Wildlife. Biodiversity Action Plans - the challenge in Wales.* CCW publication.

Andrews, E. and Crawford, A.K. (1986). *Otter Survey of Wales 1984 - 85.* The Vincent Wildlife Trust, London.

Andrews, E., Howell, P., and Johnson, K. (1993). *Otter Survey of Wales 1991.* The Vincent Wildl ife Trust, London.

Arnold, H.R. (1993). *Atlas of Mammals in Britain.* ITE Research Publication No 6, HMSO.

Barker, T.W. (1905). *Handbook to the Natural History of Carmarthenshire.* Spurrell, Carmarthen

Barratt, E.M., Deauville, R., Burland, T.M., Bruford, M.W., Jones, G., Racey, P.A. and Wayne, R.K. (1997). DNA answers the call of pipistrelle bat species. *Nature* **387**: 138-139

Barrett-Hamilton, G.E.H. (1910-11). *A History of British Mammals. 1 Bats.* Gurney and Jackson, London.

Bell, T. (1874). *A History of British Quadrupeds.* John van Boorst, London.

Berry, R.J. (1991). The house mouse. In Corbet G.B. and Harris S. (ed) *The Handbook of British Mammals* 3rd ed. Blackwell Scientific Publications.

Birks, J. (1993). The Return of the Polecat. *British Wildlife* **5**: 16-25

Birks, J.D.S. and Dunstone, N. (1991). The mink. In Corbet G.B. and Harris S. (ed) *The Handbook of British Mammals* 3rd ed. Blackwell Scientific Publications.

Blackmore, M. (1955). The Bats of Wales. *Nature in Wales* **1**: 63-67

Blandford, P.R.S. (1986). *Behavioural ecology of the polecat* Mustela putorius *in Wales.* PhD thesis University of Exeter.

Boyce, C.C.K. (1991). The water vole. In Corbet G.B. and Harris S. (ed) *The Handbook of British Mammals* 3rd ed. Blackwell Scientific Publications.

Bright, P.W. (1995). Distribution of the Dormouse *Muscardinus avellanarius* in Wales, on the edge of its range. *Mammal Review* **25**: 101-110

Bright, P. and Morris, P. (1990). *A practical guide to dormouse conservation.* Mammal Society occasional publication No. 11.

Bright, P. and Morris, P. (1993). The Conservation of the Dormouse. *British Wildlife* **4**: 154-162

Bright, P.W., Mitchell, P and Morris, P.A. (1994). Dormouse distribution: survey techniques, insular ecology and selection of sites for conservation. *Journal of Applied Ecology* **31**: 329-339

Brunker, J. (*c.* 1950). Llanegwad Parish, natural history section. Unpublished manuscript held at Carmarthen Records Office, ref. CDX/259/11.

Cadman, W.A. (1957). Vole damage in Wales, 1956-57. *Nature in Wales* **3**: 504.

Cadman,W.A. (1958). Deer in Wales. *Nature in Wales* **4**: 623

Carter, P.W. (1957). Botanical Exploration in Carmarthenshire. *The Carmarthen Antiquary* **2**: 170-171

Chanin, P. (1985). *The Natural History of Otters.* Croom Helm. London

Chapman, N.G. (1991). The chinese muntjac. In Corbet G.B. and Harris S. (ed) *The Handbook of British Mammals* 3rd ed. Blackwell Scientific Publications

Chapman, N., Harris, S. and Stanford, A. (1994). Reeves' Muntjac *Muntiacus reevesi* in Britain: their history, spread, habitat selection and the role of human intervention in accelerating their dispersal. *Mammal Review* **24**: 113-160

Churchfield, S. (1991a). The common shrew. In Corbet G.B. and Harris S. (ed) *The Handbook of British Mammals* 3rd ed. Blackwell Scientific Publications.

Churchfield, S. (1991b). The pygmy shrew. In Corbet G.B. and Harris S. (ed) *The Handbook of British Mammals* 3rd ed. Blackwell Scientific Publications.

Churchfield, S. (1991c). The water shrew. In Corbet G.B. and Harris S. (ed) *The Handbook of British Mammals* 3rd ed. Blackwell Scientific Publications.

Clarke, G.Ll. (1997). *Dormouse Survey of N.E. Carmarthenshire.* Internal CCW report.

Clegg, J. (1969). Excavations at Coygan Cave, near Laugharne. *Carmarthenshire Antiquary* **5**: 13-20

Condry, W.M. (1955). Mammal notes. *Nature in Wales* **1**: 8

Condry, W.M. (1981). *The Natural History of Wales.* Collins New Naturalist Series. London.

Congreve, A.L. (1927). Wolf in Carmarthenshire. *Transactions of the Carmarthenshire Antiquarian Society* **21**: 46

Corbet, G.B. (1971). Provisional distribution maps of British mammals. *Mammal Review* **1**: 95-139

Corbet, G.B. and Harris, S. (1991). *The Handbook of British Mammals* 3rd ed. Blackwell Scientific Publications.

Corke, D. (1977). A combination of extensive and intensive survey techniques for the study of the occurrence of *Apodemus flavicollis* in Essex. *Journal of Zoology* **182**: 171-175

Cowan, D.P. (1991). The rabbit. In Corbet G.B. and Harris S. (ed) *The Handbook of British Mammals* 3rd ed. Blackwell Scientific Publications.

Crawford, A., Evans, D., Jones, A. and McNulty, J. (1979). *Otter Survey of Wales 1977-78.* Society for the Promotion of Nature Conservation. Nettleham, Lincoln.

Cresswell, P., Harris, S. and Jeffries, D.J. (1990). *The history, distribution, status and habitat requirements of the badger in Britain.* NCC report.

Cymdeithas Edward Llwyd (1994). *Cyfres Enwau Creaduriaid a Phlanhigion :1 Creaduriaid Asgwrn-Cefn.* Gwasg y Lolfa, Talybont.

Dale-Jones, E. (1993). Three Generations: The Jeffreys of King Street, Carmarthen (1846-1936). *The Carmarthenshire Antiquary* **29**: 73-88

Davies, J. (1990). *Hanes Cymru.* Penguin. London.

Davies, M. (1968). *Ogof Pentywyn, Pendine, Carmarthenshire.* Unpublished note in CCW Carmarthenshire files.

Davies, W. (1858). *Llandeilo-Vawr and its Neighborhood.* DW and G Jones. Llandeilo.

Davis, T.A.W. (1955). Mammal notes. *Nature in Wales* **1**: 181

Davis, T.A.W. (1956). Mammal notes. *Nature in Wales* **2**: 231

Davis, T.A.W. (1957). Mammal notes. *Nature in Wales* **3**: 446

Davis, T.A.W. (1959). Mammal notes. *Nature in Wales* **5**: 865

Davis, T.A.W. (1963). Mammal notes. *Nature in Wales* **8**: 140

'Deerfoot' (1929a). Wild Nature in Llanelli. *Llanelli Mercury* 10 April 1929

'Deerfoot' (1929b). Wild Nature in Llanelli. *Llanelli Mercury* 11 April 1929

Dignam, C. (undated). *Dinefwr Park.* Unpublished report to the National Trust. Ref. PRN 25543

Dunstone, N. (1993). *The Mink.* Poyser. London.

Duverge, L. (1997). Foraging activity and habitat use of Greater Horseshoe Bats *Rhinolophus ferrumequinum* in Wales; preliminary results from Slebech Hall. The Vincent Wildlife Trust.

Evans, J. (1933). The Cryngae Hounds. *Transactions of the Carmarthen Antiquarian Society* **24**: 33

Evans, P.G.H. (1991). Chapter 9, Whales Dolphins and Porpoises: Order Cetacea. In Corbet G.B. and Harris S.(ed). *The Handbook of British Mammals* 3rd ed. Blackwell Scientific Publications.

Fenton, R. (1917 edition). *Tours in Wales 1804-13.* Bedford Press. London.

Flowerdew, J.R. (1991). The wood mouse. In Corbet G.B. and Harris S. (ed) *The Handbook of British Mammals* 3rd ed. Blackwell Scientific Publications.

Gipps, J.H.W. and Alibhai, S.K. (1991). The field vole. In Corbet G.B. and Harris S. (ed) *The Handbook of British Mammals* 3rd ed. Blackwell Scientific Publications.

Griffiths, B. and Jones, D.G. (1995). *The Welsh Academy English Welsh Dictionary.* University of Wales Press. Cardiff.

Grimes, W.F. and Cowley, L.F. (1935). Coygan Cave, Llansadyrnin, Carmarthenshire. *Archeologica Cambrensis* **90**: 95-111

Gurnell, J. (1991a). The red squirrel. In Corbet G.B. and Harris S. (ed) *The Handbook of British Mammals* 3rd ed. Blackwell Scientific Publications.

Gurnell, J. (1991b). The grey squirrel. In Corbet G.B. and Harris S. (ed) *The Handbook of British Mammals* 3rd ed. Blackwell Scientific Publications.

Gurnell, J. and Pepper, H. (1993). A critical look at conserving the British Red Squirrel *Sciurius vulgaris. Mammal Review* **23**: 127-137

Harris, S. and Lloyd, H.G. (1991). The fox. In Corbet G.B. and Harris S. (ed) *The Handbook of British Mammals* 3rd ed. Blackwell Scientific Publications.

Harris, S., Morris, P.A., Wray, S. and Yalden, D.W. (1993). *A review of British mammals: population estimates and conservation status of British mammals other than cetaceans.* JNCC report. Peterborough.

Harting, J.E. (1893). Notes on the Pine Marten. *The Zoologist* **15**: 401-409

Hill, G.R. (1961). Rats in Wales. *Nature in Wales* **7**: 43-45

Hill, G.R. (1964). Wild Mink in Wales. *Nature in Wales* **9**: 17-18

Hills, D. (1991). Ephemeral Introductions and Escapes. In Corbet G.B. and Harris S. (ed) *The Handbook of British Mammals* 3rd ed. Blackwell Scientific Publications.

Holiday, F.W. (1961). When Mink Go Wild. *The Field* 5 Oct. 1961 p609

Hope Jones, P. (1974). Welsh Whales. *Nature in Wales* **14**: 115-116

Howard, R.W. (1995). *Auritus: a natural history of the brown long-eared bat.* William Sessions. York.

Howes, C.A. (1979). The noctule bat *Nyctalus noctula* in Yorkshire. *Naturalist* **XX**: 107-110

Humphries, J. (1971). The Gwenffwrd Reserve. *Nature in Wales* **12**: 250-255

Hurrell, E. and McIntosh, G. (1984). Mammal Society dormouse survey, January 1975 - April 1979. *Mammal Review* **14**: 1-18

Hutson, A.M. (1985). Whiskered bats in south east Britain. *Bat News* **11**: 2-3

Hutson, A.M. (1997). Two species of bat new to the UK. *Bat News* **46**: 2-3

James, D.B. (undated). *Myddfai: its Land and Peoples.*

James, E.O. (1931). Letter dated 6.4.31. NMW files.

James, E.O. (1959). Letter to Colin Matheson dated 13.3.59. NMW files.

James, H. (in prep.). *The History and Archaeology of the Rabbit in Wales.*

James, T. (1991). Where Sea Meets Land. The Changing Carmarthenshire Coastline. *Sir Gar: Studies in Carmarthenshire History* **4**: 143-165. Carmarthen Antiquarian Society.

Jenkins, D. (1990). *The Law of Hywel Dda.* Gomer Press.

Jones, F. (1987). The Families of Talycrwth and Pant Howel in the Parish of Trelech a'r Betws. *Carmarthenshire Antiquary* **23**: 29-39

King, C.M. (1991a). The stoat. In Corbet G.B. and Harris S. (ed) *The Handbook of British Mammals* 3rd ed. Blackwell Scientific Publications.

King, C.M. (1991b). The weasel. In Corbet G.B. and Harris S. (ed) *The Handbook of British Mammals* 3rd ed. Blackwell Scientific Publications

King, G.M. (1930). Otters and Badgers in Carmarthenshire. *Transactions of the Carmarthen Antiquarian Society* **22**: 63-65

Langley, P.J.W. and Yalden, D.W. (1977). The decline of the rare carnivores in Great Britain during the nineteenth century. *Mammal Review* **7**: 95-116

Lever, C. (1977). *The Naturalised Animals of the British Isles.* Hutchinson and Co. London.

Linnard, W. (1982). *Welsh Woods and Forests: History and Utilization.* National Museum of Wales, Cardiff.

Linnard, W. (1984). Hurdling the Roe. *Deer* **6**: 173

Lister, J. and Whitbread, A. (1988). *Carmarthenshire Inventory of Ancient Woodland (Provisional).* NCC report.

Lloyd, H.G. (1980). *The Red Fox.* Batsford. London.

Lloyd-James, J. (1909). Traces of the Romans in the Taf Valley and its Vicinities. *Transactions of the Carmarthen Antiquarian Society* **4**: 78-87

Lockley, R.M. (1970). *The Naturalist in Wales.* David and Charles, Newton Abbott.

Lockley, R.M. (1962). The Grey Squirrel Invasion in 1961. *Nature in Wales* **8**: 36

Lovegrove R., Williams, G. and Williams, I. (1994). *Birds in Wales.* Poyser London.

Lucas, A. (1993a). Dormice in Carmarthenshire - filling in the gaps. *Llanelli Naturalists Magazine* Winter 1992-1993, 30-31

Lucas, A. (1993b). Dormice in Wales. *British Wildlife* **4**: 340

Lucas, A. (1994). Carmarthen Bat Group Report 1993. *Llanelli Naturalists Newsletter* Winter 1993-1994, 26-28

Lucas, A. (1995). Carmarthen Bat Group Report 1994. *Llanelli Naturalists Newsletter* Summer 1995, 5-7.

Lucas, A. (1996). Carmarthen Mammal Report 1995. *Bulletin of the Llanelli Naturalists* **2** 7-9

Matheson, C. (1931). *The Brown and Black Rat in Wales.* National Museum of Wales.

Matheson, C. (1932). *Changes in the Fauna of Wales Within Historic Times.* National Museum of Wales.

Matheson, C. (1941). The Rabbit and Hare in Wales. *Antiquity* **15**: 371-81

Matheson, C. (1950). Letter to Capt. H.R.H. Vaughan dated 16.12.50. NMW files.

Matheson, C. (1955). The Carnivores of Wales. *Nature in Wales* **1**: 22-24

Matheson, C. (1955). The Rodents of Wales. *Nature in Wales* **1**: 111-115

Matheson, C. (1965). The distribution of the Yellow-Necked Mouse *Apodemus flavicollis* in Wales and its border counties. *Nature in Wales* **9**: 141-143.

Matheson, C. (1966). Letter to Forestry Commission, 21.10.66. NMW files.

Matheson, C. and Cowley, L.F. (1934). *Pseudorca crassidens* (Owen) on the Glamorgan Coast. *Nature* **133**: 870.

McDonald, R., Bright, P.W. and Harris, S. (1994). *Baseline Survey of Pine Martens in Wales.* CCW Contract Science Report 103.

McOwat, T. (1994). Dyfed Bat Group Winter Counts 1993-94 Unpublished report.

Mathew, N.R. (1991a). Some Observations on the Otter (*Lutra lutra*) on the River Tywi at Llandeilo. *Llanelli Naturalists Magazine* Summer 1992, 23-40.

Mathew, N.R. (1991b). Bats in Carmarthenshire. *Llanelli Naturalists Newsletter* Winter 1990-1991, 18-21.

Mathew, N.R. (1992). The Dormouse. *Llanelli Naturalists Magazine* Winter 1991-1992, 9-10.

Mathew, N.R. (1993). A survey of the distribution of the dormouse in SE Carmarthenshire. *Llanelli Naturalists Magazine* Summer 1993, 10-22.

Mathew, N.R. (1997). *The Dormouse in Carmarthenshire.* BSc. Dissertation, Trinity College, Carmarthen.

Mee (1892). Notes of a visit to Carmarthenshire in 1687 by John, First Lord Ashburnham. *Carmarthen Miscellany 1*

Messenger, J., Birks, J. and Jeffries, D. (1997). What is the Status of the Pine Marten in England and Wales? *British Wildlife* **8**: 273-279

Millias, J.G. (1904-6). *The Mammals of Great Britain and Ireland.* Longman Green. London.

Mitchell-Jones, A.J. (1990). The distribution of bats in Britain 1982-87 as revealed by enquiries. *Mammal Review* **20**: 145-157.

Montgomery, W.I. (1991). The yellow-necked mouse. In Corbet G.B. and Harris S. (ed) *The Handbook of British Mammals* 3rd ed. Blackwell Scientific Publications.

Morgan, I K. (1982). John Brunker - Llanegwad Naturalist - His notes and records. *Llanelli Naturalists Newsletter* Summer 1982, 11-15.

Morgan, I.K. (1993). Interim Notes on the Status of the Pine Marten in South and Mid Wales. *Llanelli Naturalists Newsletter* Winter 1992-93, 11-22.

Morris, J.E. (1901). *The Welsh Wars of Edward I.* Clarendon Press. Oxford.

Morris, P.A. (1983). *Hedgehogs.* Whittet Books. London.

Morris, P.A. (1993). *A Red Data Book For British Mammals.* Mammal Society.

Morris, P.A. (1995). Speedy hedgehogs and media myths. *New Scientist* 14 January 1995, 44-45.

Morris P.A. (1991). The hedgehog. In Corbet G.B. and Harris S. (ed) *The Handbook of British Mammals* 3rd ed. Blackwell Scientific Publications.

Morris, E.H. and Parkinson, T. (1897). *Circulars to the members of the Carmarthen Hunt.* Correspondence kept at Carmarthenshire Archive Office, Box Bryn Myrddin 158.

Neal, E.G. (1986). *The Natural History of Badgers.* Croom Helm. Beckenham.

Nicholson, J.A. (1996). *Further Historical Glimpses: Pembrey and Burry Port.* Book 3. Carmarthenshire County Council.

Pain, S. (1997). Out of the Woods. *New Scientist,* **2096**: 10

Pease, A.E. (1898). *The badger.* Lawrence and Bullen. London.

Panting, P.J. (1971). Mammal field notes. *Nature in Wales* **12**: 181-183

Panting, P.J. (1972). Mammal field notes. *Nature in Wales* **13**: 55

Panting, P.J. (1973). Mammal field notes. *Nature in Wales* **13**: 207-208

Perrow, M. and Jowitt, A. (1995). What Future for the Harvest Mouse? *British Wildlife* **6**: 356-365.

Phillips, L. (1885). Dormouse in Cardiganshire. *The Zoologist* **9**: 258

Pocock R.I. (1927). Some Interesting Mediaeval Animal Relics. *Transactions of the Carmarthenshire Antiquarian Society* **21**: 46-47.

Pryce, R. (1994). Twenty-one Years of the Llanelli Naturalists. *Llanelli Naturalists Newsletter* Winter 1993-94, 5-14.

Ransome, R.D. (1990). *The Natural History of Hibernating Bats.* Christopher Helm. London.

Ransome, R.D. (1991a). The greater horseshoe bat. In Corbet G.B. and Harris S. (ed) *The Handbook of British Mammals* 3rd ed. Blackwell Scientific Publications.

Ransome, R.D. (1991b). The lesser horseshoe bat. In Corbet G.B. and Harris S. (ed) *The Handbook of British Mammals* 3rd ed. Blackwell Scientific Publications.

Ransome, R.D. and McOwat, T.P. (1994). Birth timing and population changes in greater horseshoe bat colonies *(Rhinolophus ferrumequinum)* are synchronised by climatic temperature. *Zoological Journal of the Linnean Society* **112**: 337-351.

Rees, D. (1995). The Forest of Glyncothi. *The Carmarthenshire Antiquary* **31**: 45-55.

Rees, W. (1953). *A Survey of the Dutchy of Lancaster Lordships in Wales 1609-1613.* University of Wales Press. Cardiff.

Roese, H.E. (1979). Archaeological Discoveries by T.C. Cantrill. *Archeologica Cambrensis* **128**: 147-155

Russell, W.D. (1950). Letter dated 4.12.50. NMW files.

Samuel, W. (1868). *Llandilo Present and Past.*

Sansbury, D.G. (1956). Mammal Notes. *Nature in Wales* **3**: 404-405

Schober, W, and Grimmberger, E. (1989). *A Guide to the Bats of Britain and Europe.* Hamlyn Publishing Group.

Slater, F.M. (1987). Mammal notes. *Nature in Wales* **6** (new series): 76.

Slow, W.J. (1950). Letter dated 15.11.50. NMW files.

Speakman, J.R. (1991). The Daubenton's bat. In Corbet G.B. and Harris S. (ed) *The Handbook of British Mammals* 3rd ed. Blackwell Scientific Publications.

Stebbings, R.E. (1966). A population study of bats of the genus *Plecotus*. *Journal of Zoology* **150**: 53-75.

Stone, R.D. and Gorman, M.L. (1991). The mole. In Corbet G.B. and Harris S. (ed) *The Handbook of British Mammals* 3rd ed. Blackwell Scientific Publications.

Strachan, R and Jefferies, D J (1993). *The Water Vole (Arvicola terrestris) in Britain 1989-90: its distribution and changing status.* The Vincent Wildlife Trust

Staines, B.W. and Ratcliffe, P.R. (1991). The roe deer. In Corbet G.B. and Harris S. (ed) *The Handbook of British Mammals* 3rd ed. Blackwell Scientific Publications.

Stebbings, R.E. (1991). The Brandt's bat. In Corbet G.B. and Harris S. (ed) *The Handbook of British Mammals* 3rd ed. Blackwell Scientific Publications.

Stephens, M.N. (1957). *The Otter Report.* Universities Federation for Animal Welfare, London.

Swift, S.M. and Racey, P.A. (1983). Resource partitioning in two species of vespertilionid bats (Chiroptera) occupying the same roost. *Journal of Zoology* **200**: 249-259

Tapper, S.C. (1992). *Game Heritage.* Game Conservancy Ltd.

Tapper, S.C. (1991). The brown hare. In Corbet G.B. and Harris S. (ed) *The Handbook of British Mammals* 3rd ed .Blackwell Scientific Publications.

Tapper, S.C. and Parsons, N. (1984). The changing status of the brown hare (*Lepus capensis* L.) in Britain. *Mammal Review* **14**: 57-70.

Taylor, K.D., Fenn, M.G. and Macdonald, D.W. (1991). The brown rat. In Corbet G.B. and Harris S. (ed) *The Handbook of British Mammals* 3rd ed. Blackwell Scientific Publications.

Thomas, J.F. (1955). The Laugharne Pendine Burrows. *Nature in Wales* **1**: 72-74

Thompson, H.V. (1964). Wild Mink. *Agriculture* **26**: 564-567

Thompson, H.V. and Worden, A.N. (1956). *The Rabbit.* Collins New Naturalist.

Thorpe, L. (1978). *Gerald of Wales. The Journey Through Wales/The Description of Wales.* Penguin Classics. London.

Twigg, G.I. (1984). *The Black Death.* Batsford. London.

Turner-Evans, H. (1962). Letter to Colin Matheson dated 4.10.62. NMW files.

Vaughan, H.R.H. (1950). Letter to Colin Matheson dated 10.12.50. NMW files.

Velander, K.A. (1983). *Pine Marten Survey of Scotland, England and Wales, 1980-1982.* The Vincent Wildlife Trust, London.

Vesey-Fitzgerald, B. (1949). *British Bats.* Methuen. London.

Walton K.C. (1968). The baculum as an age indicator in the polecat *Putorius putorius. Journal of Zoology* **156**: 533-536.

Watts, C.H.S. (1968). The foods eaten by wood mice *Apodemus sylvaticus* and bank voles *Clethrionomys glareolus* in Wytham Woods, Berkshire. *Journal of Animal Ecology* **161**: 25-41

Williams, G.A. (1985). *When Was Wales.* Penguin, Reading.

Woodroffe, G. (1994). The Water Vole - some aspects of its ecology. *British Wildlife* **5**: 296-303.

Wright, F.S. (1916). The mammalian fauna of North Cardiganshire, with notes on certain peculiarities and rare species. *Zoologist* Ser. **4** : 20, 329

Yalden, D.W. (1981). The occurrence of the pygmy shrew *Sorex minutus* on moorland and the implications for its presence in Ireland. *Journal of Zoology* **195**: 81-92.